# Mapping Work Processes

## Second Edition

Also available from ASQ Quality Press:

*The Certified Quality Process Analyst Handbook*
Eldon H. Christensen, Kathleen M. Coombes-Betz, and Marilyn S. Stein

*Enterprise Process Mapping: Integrating Systems for Compliance and Business Excellence*
Charles G. Cobb

*Business Process Improvement Toolbox,* Second Edition
Bjørn Andersen

*The Process-Focused Organization: A Transition Strategy for Success*
Robert A. Gardner

*Root Cause Analysis: Simplified Tools and Techniques,* Second Edition
Bjørn Andersen and Tom Fagerhaug

*Managing Service Delivery Processes: Linking Strategy to Operations*
Jean Harvey

*The Process Auditing Techniques Guide*
J.P. Russell

*Defining and Analyzing a Business Process: A Six Sigma Pocket Guide*
Jeffrey N. Lowenthal

*The Logical Thinking Process: A Systems Approach to Complex Problem Solving*
H. William Dettmer

*Enabling Excellence: The Seven Elements Essential to Achieving Competitive Advantage*
Timothy A. Pine

*Lean Kaizen: A Simplified Approach to Process Improvements*
George Alukal and Anthony Manos

*The Certified Manager of Quality/Organizational Excellence Handbook,* Third Edition
Russell T. Westcott, editor

*Performance Measurement Explained: Designing and Implementing Your State-of-the-Art System*
Tom Fagerhaug and Bjørn Andersen

*Bringing Business Ethics to Life: Achieving Corporate Social Responsibility*
Bjørn Andersen

To request a complimentary catalog of ASQ Quality Press publications,
call 800-248-1946, or visit our Web site at http://www.asq.org/quality-press.

# Mapping Work Processes

## Second Edition

Bjørn Andersen
Tom Fagerhaug
Bjørnar Henriksen
Lars E. Onsøyen

ASQ Quality Press
Milwaukee, Wisconsin

American Society for Quality, Quality Press, Milwaukee 53203
© 2008 by ASQ
All rights reserved. Published 2008
Printed in the United States of America
23 22 21 20 19     8 7 6 5 4

Library of Congress Cataloging-in-Publication Data

Mapping work processes / Bjørn Andersen . . . [et al.].—2nd ed.
    p. cm.
  Rev. ed. of: Mapping work processes / Dianne Galloway. c1994.
  Includes bibliographical references and index.
  ISBN 978-0-87389-687-0 (soft cover, spiral bound : alk. paper)
  1. Work design.  2. Flow charts.  I. Andersen, Bjørn.  II. Galloway, Dianne,
 1938– Mapping work processes.

  T60.8.G35 2008
  658.5'4—dc22                                                2008009079

ISBN: 978-0-87389-687-0

Publisher: William A. Tony
Acquisitions Editor: Matt T. Meinholz
Project Editor: Paul O'Mara
Production Administrator: Randall Benson

ASQ Mission: The American Society for Quality advances individual, organizational, and community excellence worldwide through learning, quality improvement, and knowledge exchange.

Attention Bookstores, Wholesalers, Schools, and Corporations: ASQ Quality Press books, videotapes, audiotapes, and software are available at quantity discounts with bulk purchases for business, educational, or instructional use. For information, please contact ASQ Quality Press at 800-248-1946, or write to ASQ Quality Press, P.O. Box 3005, Milwaukee, WI 53201-3005.

To place orders or to request a free copy of the ASQ Quality Press Publications Catalog, including ASQ membership information, call 800-248-1946. Visit our Web site at www.asq.org or http://www.asq.org/quality-press.

Printed in the United States of America

∞ Printed on acid-free paper

Quality Press
600 N. Plankinton Avenue
Milwaukee, Wisconsin 53203
Call toll free 800-248-1946
Fax 414-272-1734
www.asq.org
http://www.asq.org/quality-press
http://standardsgroup.asq.org
E-mail: authors@asq.org

AMERICAN SOCIETY
FOR QUALITY®

# Table of Contents

# List of Figures and Tables

# Preface

As a reader of this book, to you it might not be apparent that for the authors, this second edition is indeed quite unusual. Unusual in the sense that we have done a new version of a book that Dianne Galloway wrote originally. None of us know her, and have never met her, but we've known of her book since it was published. It has been a strong bestseller for ASQ Quality Press ever since it was first published, with more than 40,000 copies in circulation, and we've found the book very instructive and useful.

Being more than ten years old, the book did, however, show signs of aging and needed a revision. When Ms. Galloway declined to undertake this task, we were asked whether we'd be interested in taking it on. Slightly awestruck by trying to follow her act, we thought about it for a few days, decided it was possible, and forged ahead.

You might wonder what our credentials are for embarking on such a tough task. All of us having worked in the field of process improvement for 10 to 15 years, we believe we have accumulated a fair amount of experience and insight. Throughout the years, we've worked on projects to map and improve work processes in a wide variety of industries and types of organizations, from traditional manufacturing companies to service providers, banks, the telecom sector, public sector organizations like municipalities and state institutions, hospitals, energy companies, and others. Supplemented by numerous research projects and studies on the topic of work process analysis and improvement, process-oriented organizational models, and related areas, we have seen quite a few facets of work process mapping.

This might be a suitable time to define what we mean by a work process. The main component in this term is the word *process*, which in its most

*Definition of work process*

basic form can be defined as ". . . a logical series of related transactions that converts input to results or output." To separate a company's processes from other types of processes, the word *work* (or equally often used, *business*) has been added to form the term *work process*. A work process can be defined in a number of different ways, for example:

> *A chain of logically connected, repetitive activities that utilizes the organization's resources to refine an object (physical or mental) for the purpose of achieving specified and measurable results/products for internal or external customers.*

In a similar vein, business process reengineering pioneers Hammer and Champy defined a business process as ". . . *a collection of activities that takes one or more kinds of input and creates an output that is of value to the customer.*" These different definitions all help clarify what we mean by work processes, and we don't subscribe to any one in particular.

Like the first edition, our purpose in writing this book has been to provide a practical, introductory text on how to map work processes. There are many articles and books that discuss and explain more advanced approaches and applications of process maps (and we encourage our readers to move on to these after completing this book) but not many that provide first-timers or novice process mappers with step-by-step instructions. The overwhelming popularity of the first edition clearly shows that there is a need for such a text, thus we have strived to keep it that way while bringing it up to date on developments that have taken place over the last decade.

*Changes from first edition*

If you are familiar with the first edition, you will find that the most noticeable change is the inclusion of several new types of process maps. At the time the first edition was written, the basic, straightforward flowchart was, if not the only one around, by far the most widely used type of process map. In 2008, the basic flowchart is still extensively used, but has been supplemented by a number of other types, all of which serve different purposes. This means that less space is devoted to the art of developing the details of a basic flowchart, but we make up for this by showing you a wider variety of charts you can use (and you can always reference both editions, reviewing the very detailed explanations in the first, if necessary).

*Structure of the book*

The book is structured as follows. Chapter 1 presents an introductory explanation of why mapping work processes is an important activity in an organization (and as pointed out by Galloway, both as an important process in itself and as a means to other ends). In Chapter 2, we outline the different types of process maps available, while Chapter 3 describes the *meta-process* of mapping a work process, including roles, overall steps, and many pieces of practical advice. In Chapters 4 through 8, five different process maps are explained in detail. Within each chapter, we describe the type of process map, lay out the steps in constructing one, provide hints and advice that we have acquired the hard way through learning and failing, illustrate the type of map with a case study, and close each chapter with a checklist. We should also point out that if you are already familiar with some of the different types of process maps, these chapters are designed to stand alone, allowing you to read just the ones you'd like to learn more about. Finally, Chapter 9 rounds off the book by discussing briefly how you can use the process maps for different purposes. We have also provided a small glossary at the end to ensure that all terms used throughout the book are understood.

Hopefully, you will notice that we have strived to make the layout and visual appearance of the book inviting for practitioners and readers who quickly want to grasp the approaches described. We have deliberately omitted academic references from the text, and the annotations in the margins are meant to help you navigate the text easily. We have sprinkled the book with ample examples from many different industries, and we have favored an "airy" layout with plenty of illustrations over lengthy explanations and discussions. Should you find that you need more in-depth treatments of various topics, just consult more academically oriented books and we are confident you will be able to find what you need.

There is a case study used to illustrate the construction and use of the different types of process maps running through the book, introduced in Chapter 2 and developed in detail in Chapters 4 through 8. The case organization is Brook Regional Hospital, a fictitious hospital (but based on experiences from our work with different hospitals). Deciding to use a hospital as the case study was no easy decision; we debated long over whether to use a manufacturing company instead, or perhaps a typical service company. When we finally decided on a hospital, this was partly because hospitals lend themselves easily to depicting different process flows, partly because most people can relate to and recognize hospital processes, and partly because they display traits of manufacturing, service, and public sector organizations. We have worked too much with hospitals to make the mistake of saying that patients are like products that move through processing steps, but the fact is that hospitals produce both services and physical deliverables, while working closely with (or in many cases being owned by) the public sector. Thus, we think a hospital case study is quite suitable for giving a real-life view of the uses of process mapping.

We are grateful to ASQ Quality Press for giving us the opportunity to respectfully follow up Dianne Galloway's work in writing this second edition. And of course, without her writing the original, there would have been no second edition, so our warmest regards to her. There are four authors of this text, but we actually started out with six. Two of them, Ingrid Spjelkavik and Andreas R. Seim, had to pull out during the process for different reasons but have also contributed to the structure and concept of the book, for which we are truly grateful.

Bjørn Andersen
Tom Fagerhaug
Bjørnar Henriksen
Lars E. Onsøyen

Trondheim, Norway
January, 2008

# Chapter 1

# Why Map Work Processes?

From the simple fact that you are reading this book, we must infer that you are facing one of three possible situations: you are contemplating mapping a work process, you have decided to map a work process, or you have already had some experience in mapping work processes and want to learn more. (A fourth possibility is of course that someone forced you to read it—a teacher or perhaps your boss—but we'd like to focus on the first three.)

This book will describe the actual mapping process. Before embarking on this, we will put the mapping effort into a larger context. Mapping of work processes can be a useful tool in many situations, and this chapter will deal with how mapping of work processes can be both a valuable exercise and a practical tool used extensively in organizations.

Let us also point out that when mentioning work processes, many think of physical processes. Production processes in the manufacturing industry are easy to visualize. In labor-intensive work, which is most common in the knowledge industry, work processes are not as easy to see and understand. Many people may be involved in different parts of the total work process, but few may understand the whole process. In this situation, mapping creates new understanding that is important for improvement and customer satisfaction.

## 1.1 PURPOSES OF PROCESS MAPS

Depending on the extent and complexity of the process to be mapped and on the approach applied when performing the mapping, such an endeavor can be anything from a simple flowcharting task to an arduous organizational discussion involving many people. In cases where you invest significant amounts of time and resources in mapping a process, it should be comforting to know that there are numerous benefits that can be achieved from such an exercise. Furthermore, the process map itself can serve many different purposes once it has been constructed.

*The busy controller who got his weekends back*

Let us start by telling the story of a controller in a medium-sized manufacturing company we worked with some years ago. A man in his fifties, with grown-up children out of the house, he and his wife had purchased a cabin in the mountains about a three-hour drive from the town where they lived. The intention was to spend as many weekends there as possible, recovering from job stress by hiking, skiing, fishing, and so on. A financial report was due the first Monday of every month, containing about 30 pages packed with financial data, statistics, forecasts, and so on, and the controller was in charge of producing this report. On the weekends before the monthly deadline, he worked till late at night on Friday, usually a full day on Saturday, and often several hours on Sunday, so naturally those weekends were out of the question for going to the cabin. When the next weekend approached, he was usually so exhausted from the nightmare weekend before, he didn't really feel like getting in the car on Friday afternoon, fighting the traffic out of town, and arriving at the cabin late in the evening. Thus, at best, they went there a couple of weekends a month, often less when they had other plans during one or both of the remaining weekends.

We met the controller through a project to improve the purchasing processes in the company. Having mapped the core work processes performed by the purchasing department, we had moved on to look at the processes interfacing these. One of them was the process of supplying the purchasing department with financial data showing the development of prices obtained for different parts. Presto, enter our controller, whose monthly Monday report contained these data. When mapping this process, close to 15 people were involved in outlining the steps of the process and their different inputs and outputs. During the mapping exercise it soon became clear that the controller and the others differed widely in their understanding of the process. The controller drew several arrows leading out of the box labeled "financial report," sending it off to marketing, purchasing, the CEO, manufacturing, and so on. He even specified what information was required by each and where the data came from. Most of the other group members looked at him with confusion. Marketing didn't use any of the data at all, they collected their own statistics. Manufacturing picked out one number, manufacturing costs per unit, and discussed it during their first-Friday-of-the-month meeting. Once each quarter the CEO might, or might not, include a couple of charts from the report in his quarterly results presentation to the board, and purchasing used data from the report to assemble price development histories before meeting with suppliers. Typically, supplier talks were held twice each year, so there was no urgency whatsoever in purchasing obtaining the report.

The further the mapping progressed, the more it became clear that over the years, the controller had created an image in his mind that the Monday report was of pivotal importance to a number of departments throughout the company. Someone would actually need some data and ask him for the report often enough that this perception was reinforced, to the point where his cabin dream was in jeopardy of being put on hold till retirement. Realizing this, the mapping was used to clarify who needed what data at which time, effectively changing the monthly report into tailored information packages for each recipient. Some of these were needed only once per quarter, others still monthly, but not necessarily on the first Monday. Altogether, the data to be gathered and compiled into reports was reduced to about one-fifth of the old report.

This turned the controller's weekends around, freeing him of the monthly nightmare weekend. The frequency of weekends spent at the cabin increased dramatically, improving quality of life for both the controller and his wife significantly. And all this from a simple process mapping exercise! If one of us happens to run into this controller (who still has the same job) on occasion, he always lights up and insists on thanking us for giving him his weekends back.

So, what can this little story tell us (beyond the fact that making ill-founded assumptions is unwise)? We think it provides a powerful example of one of the benefits process mapping can lead to: a shared understanding among those involved in a process (or receiving outputs from it, or delivering inputs to it) of the steps in the process, its nature, and what it is intended to produce. Unnecessary work was prevented from being performed, thus eliminating much waste of time and effort. By clarifying what each internal customer really needed from the controller, the processes ran smoother, with the right inputs. These are no small feats in themselves, but are in fact just some of the many positive outcomes of mapping work processes (and many other types of process modeling as well):

- Process maps are visual depictions of the steps and flows of a process, and such pictures make understanding and communication about the process much easier.

- Process modeling requires confrontation and clarification where disagreements or ambiguities are present, and does not allow vague assumptions.

*The benefits of mapping work processes*

- Process mapping leads to a general increase in the level of specification; assumptions, the environment of the process, both physical and intangible, and objectives must be explicitly expressed.

- Mapping a work process means tapping the knowledge of everyone who is or has been involved in designing and running that process; this is an invaluable heirloom of the organization, often containing years if not decades of hard-earned experience.

- A model also allows more advanced simulations or scenario testing and can be used as a tool in what-if assessments.

- A process map provides documentation of a deliberately designed process. It can be used to demonstrate to customers, regulatory bodies, or other stakeholders, that the organization employs a systematic approach to process design and fulfills mandated requirements. For customers, especially industrial customers or potential industrial customers, the model can be used as a marketing tool to convince them that your business processes are sound and reliable.

- Many types of standards and certifications require extensive process mapping, including the process-based year 2000 version of ISO 9000.

- If done right, that is, in collaboration with many or all of those involved in the work process being mapped, it stimulates involvement and enthusiasm.

- The modeling and dialog connected with the mapping process creates knowledge and better understanding of the process and its boundaries. This might be the most important benefit; after an extensive work process mapping effort in an organization, everyone is much more aware of their role in the greater scheme of things and how their effort contributes to the overall objectives of the organization.

*Process maps have many purposes and hold many benefits*

Process maps, both while creating them and after their completion, serve many different uses in an organization, for example, for training new employees or individuals new to a certain job, identifying areas in need of improvement, creating interfaces between your process and that of a supplier or customer, or marketing your high-quality processes. We will continue this discussion in more detail in Chapter 9, after we have presented some of the different types of process maps available.

A word of caution is also in order: while we hail process maps as useful tools in many different contexts of an organization, we must point out that the process of creating the map is equally, if not more, important. It is the shared experience of a group working out the steps of the process, how they are sequenced, what they produce, and generally ironing out the kinks in the process that creates a mutual understanding of it. It is the mapping exercise that brings people together around a table who otherwise never discuss the process, and that provides an opportunity for discussion and communication. Our message is certainly not that the process map is irrelevant; as we have mentioned already, it serves many different purposes. But it is not so critical a goal that it justifies taking shortcuts or commissioning a couple of persons to construct the map just to get it done. A successful *process mapping* exercise is the product of a well-designed *mapping process* and results in a high-quality *process map*. This also reflects the organization of this book, with Chapter 3 focusing on how to organize and execute the process of mapping work processes, and Chapters 4 through 8 focusing on different types of maps.

## 1.2 WHAT ARE WORK PROCESSES?

We promise not to bore you with numerous instances of definitions throughout this book; we know they do not make for interesting reading. And while we intend to keep that promise, we would be negligent in the execution of our duties if we did not introduce the very topic of process and the terms we will use in this book. First of all, you no doubt either have heard or are already familiar with terms like *work process*, *business process*, *value-creating process*, *primary process*, and probably a few others in the same vein. If you look up the word *process* in a dictionary, it will say something like this:

> . . . *a logical series of related transactions that converts input to results or output.*

*Definitions of work processes*

This being very generic, people have added the modifier *work* or *business* to make it clear they are talking about processes that take place inside an organization, as part of work or business. A large company trying to clarify these terms to their employees early in the 1990s defined such processes as:

> *A chain of logically connected, repetitive activities that utilizes the organization's resources to refine an object (physical or mental) for*

*the purpose of achieving specified and measurable results/products for internal or external customers.*

This definition works for us; the approaches described in this book are aimed at any process being performed in an organization, be it a short, simple task or a complex sequence of events running across departments or even organizational boundaries. Two other terms mentioned above are used to classify work processes. Many label the processes that actually deliver a service or product to customers *primary processes* or *value-creating processes*, for example the payment process in a grocery store, the hair cutting process in a salon, the surgery process in a hospital, or the manufacturing of a TV in an electronics company. Such labels separate them from processes that are needed to allow these primary processes to run, but are not directly involved in satisfying customers, for example, recruiting employees, maintaining equipment or vehicles, or improving the supplier base. These are termed *secondary processes* or *support processes*. Some even claim that the last example, improving the supplier base, belongs to a separate group called *development processes*. If you want to study these classifications of processes or process frameworks in more detail, there are several sources listed at the end of the book.

## 1.3 THE CASE STUDY COMPANY: BROOK REGIONAL HOSPITAL

Brook Regional Hospital (BRH) was established in 1953, and has operated as a community hospital for more than 50 years, owned and run by a local women's aid society. The hospital has 254 beds, and with more than 1100 employees and a medical staff of over 250 independent practitioners, the hospital is one of the largest employers and a key institution in the community.

Founded on a philosophy of and passion for meeting community needs, BRH has been known for always striving to provide the best possible care to the local community. It is considered a good hospital in many respects, but has not been in the forefront in terms of technology investments, modern organizational principles, or expansion of areas of care.

BRH is an acute-care community hospital with extensive inpatient and outpatient services that provide care to the host community and the rest of the region. The hospital offers specialized services in oncology, pediatrics, diabetes care and treatment, orthopedics, cardiac care, rehabilitation, and pain management. These different services are run within different specialized departments located in their own physical areas, distributed over

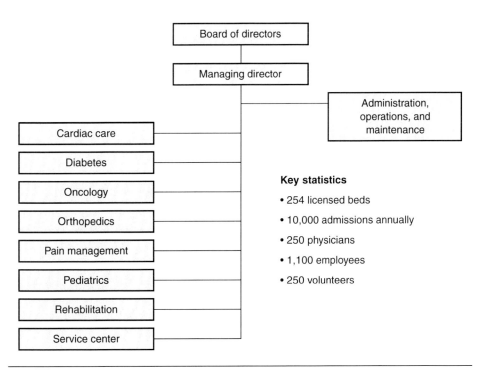

**Figure 1.1** Simplified organizational chart for BRH.

eight floors and three wings. In addition to these specialized services, the hospital has a service center offering diagnostics, X-ray, and anesthesia to the specialized services. There is also an operational and administrative unit responsible for finances, purchasing, maintenance of buildings and equipment, catering, laundry, and so on. The organizational structure is shown in Figure 1.1.

Just a few weeks ago, following months of undisclosed negotiations, it was announced that a large hospital group, People Care Hospitals (PCH), will acquire BRH from the women's aid society. The aid society has weakened over the years, finding it difficult to recruit members and volunteers, and selling the hospital will free up significant capital that can be used in other projects. PCH has successfully acquired more than two dozen regional hospitals during the last 10 years, and has grown into a powerful chain of both general care and specialized units. Being a commercial organization, they're known for demanding efficient operations and profitability, and the management of BRH is uncertain about what the future will bring.

In this context, the need arises to create a clearer understanding of the hospital, its role in the community and the PCH chain, and its work processes. Throughout the rest of the book, BRH will be used to illustrate how different approaches to mapping work processes are accomplished in practice.

# Chapter 2

---

# Different Approaches to Mapping Work Processes

Since the publication of the first edition of this book, process mapping has evolved significantly, not in the least by refining various types of process maps that already existed in 1997, but also by introducing new ways of mapping processes and adapting approaches from other fields. A more extensive arsenal of mapping methods makes process mapping an even more useful tool and extends the application areas where process mapping can offer valuable help. Thus, we have decided to deviate from the first edition and expand the number of process maps covered in the book.

## 2.1 WHEN IS IT USEFUL TO MAP WORK PROCESSES?

Instead of starting by listing the different types of process maps that exist, and then trying to convince you that each of them fills a certain need or application, we think it is more logical to first outline typical situations in which you might find it useful to map one or more processes:

- A process in your organization is being performed regularly and to standard, but no one has a full picture of the total process. Simply to understand its steps and flow of activities, there is a desire to map the process.

- New employees or people transferred from other areas of the organization need to be trained for a process. Since no process descriptions exist or existing ones are outdated, the process flow must be mapped as a platform for this training.

*Situations where there might be a need for mapping one or more work processes*

9

- The organization faces problems, for example, loss of revenue, problems in recruiting and keeping qualified employees, or increasing costs, and is not sure where to start improvements to rectify the situation. Mapping a broad range of work processes to identify areas of substandard performance can direct the improvement efforts to processes with the best payback.

- A targeted improvement project has been launched to improve the performance of a certain work process. Undertaking such a project without mapping the process is almost unthinkable, unless the problem is extremely well understood from the outset. Depending on what you suspect is causing the problems and how well you know the process, there might be a need to analyze the cost breakdown of the process, examine whether there are overlaps or gray areas in terms of responsibilities, whether bottlenecks slow down the process, whether several parallel routes through the process exist, possibly with unbalanced capacity or capabilities in the different paths, and so on.

- You are going through a cycle of updating strategic targets and plans and need to increase your understanding of the competitive situation in your market segment, your general surroundings, and the different internal and external actors that shape your future success.

- You are constructing new infrastructure, perhaps an office building, a plant, or a new hospital wing, and you see this as an opportunity to optimize less-than-ideal processes. In such a case, mapping the ideal processes to allow for designing the new infrastructure around these is a fruitful approach.

- The same applies if you are investing in a new computer system of some complexity, especially if the system allows some customization to each buyer or the software is designed from scratch for the organization. Simply automating old work processes, especially if part or all of these have previously been performed outside computer systems, is rarely a good idea. Mapping the ideal way to perform these processes as a basis for the software design is a much better approach.

- The organization faces external demands for processes to be mapped, for example, from customers that require their suppliers to clean up their processes and document their practices. In many cases, authorities or regulating bodies require organizations that want to operate in a specific sector to map their processes. If you are obtaining certification like ISO 9001, ISO 14001, variants of these, or accreditation as an operator, for example in the healthcare sector, you will normally have to provide extensive process documentation to the certifying body.

- Having realized a need for understanding better how the entire supply chain you are a part of works, the links to both first-tier suppliers and customers, as well as the full span of the chain, must be mapped.

- When two organizations are merging, either through friendly negotiations or as a result of acquisition, two organizational cultures and sets of processes are often combined to keep the best from both. This would be a near-impossible task without knowing how processes in both organizations are carried out, and at what performance levels.

Although it would be extremely nice if one type of process map could work in all these different situations, the reality is that there is no such type of wonder map. To make it easier to decide which type of map to choose in a given situation, Table 2.1 lists typical uses and the preferred type(s) of map for each (although we do not claim this to be an exhaustive list of application

**Table 2.1**  Application areas and preferred types of process maps.

| Process map use | Suitable type of process map |
|---|---|
| Obtain a general understanding of the flow of process steps | • Basic flowchart<br>• For complex processes, flowchart divided into segments or levels |
| Prepare training materials | • Basic flowchart<br>• Cross-functional flowchart<br>• For complex processes, flowchart divided into segments or levels |
| Identify areas in need of improvement | • High-level process overview map<br>• Cross-functional flowchart |
| Analyze the cost structure of a process | • Flowchart with cost data |
| Analyze time performance and capacity/ capability issues in a process | • Bottleneck map<br>• Flowchart with load statistics |
| Analyze responsibility issues in a process | • Cross-functional flowchart |
| Understand the organization's place in its surroundings and links to other actors | • Stakeholder map |
| Map ideal processes in preparation for acquiring new infrastructure or software system | • High-level process overview map<br>• Basic flowchart<br>• Cross-functional flowchart<br>• For complex processes, flowchart divided into segments or levels |

*Types of process maps suitable for different uses*

*Continued*

**Table 2.1**   Application areas and preferred types of process maps. *Continued*

| Process map use | Suitable type of process map |
|---|---|
| Map work processes to meet external requirements for process documentation | • Depends on the format required |
| Create better insight into the structure and processes of the supply chain the organization is part of | • Stakeholder map<br>• Supply chain model |
| Prepare for a merger | • High-level process overview map<br>• Basic flowchart<br>• Cross-functional flowchart<br>• For complex processes, flowchart divided into segments or levels |

areas for process maps). In later chapters we will show examples of how the different types of maps are used for various purposes, but selecting a map can be very much a matter of personal taste.

## 2.2 TYPES OF PROCESS MAPS

From the right-hand column of the table, it's apparent that process maps come in many shapes and forms, even some other types not listed here. The ones included in the table are, however, those that will be presented in this book. Each will be covered in separate chapters, containing a description of the type of map, a brief outline of the steps in constructing such a map, an example of its use in Brook Regional Hospital, and some practical advice on how to maximize the usefulness of the map. Before moving on to these detailed presentations, let us conclude this chapter by quickly describing the types of process maps included:

*The types of process maps presented in the book*

- The *stakeholder map* is a graphical representation of a stakeholder analysis. It is used to portray how the organization is located in a larger set of surroundings and what links it has to different actors, internal and external.

- The *value chain map/high-level process model* defines the organization's position in the larger chain of supplier–customer relationships that form the chain, from raw material suppliers to end vendors of products or services to the final customer, and which individual work processes are performed inside the organization and how these are linked together in chains or networks.

- The *basic flowchart* is a simple map depicting the activities, with inputs and outputs, performed in an individual process.

- The *cross-functional flowchart* uses "swim lanes" to supplement the basic flowchart with information about who is responsible for each activity, thus allowing the identification of illogical assignments of responsibility.

- The *bottleneck map* and *process map with load statistics* are two variants aimed more at process analysis, illustrating where available capacity and current demand do not match, or showing how a process can have several parallel paths through it and how the load is distributed over these paths.

So far, we have referred to work processes as if they are homogenous. The fact is that work processes vary immensely in terms of complexity, the number of steps they span, how many people are involved in executing them, and not in the least the degree of standardization and number of times they are repeated. This has implications for mapping them; standardized processes repeated frequently are more predictable and usually easier to map (and improvements in them resulting from the mapping will often have a large impact due to the volume of the process). More complex processes that last longer and involve more people, with less predictable flow, for example, innovation processes, employee development, or building a house, are often harder to map. On the other hand, creating a shared understanding of roles and flows in such processes can have a higher impact on performance since these often are more expensive processes to run and with higher impact on the future development of the organization.

# Chapter 3

---

# The Process of Mapping
a Process

As shown in Chapter 2, different types of process maps serve different purposes. This is highlighted in Chapters 4 through 8, where we describe the detailed steps you should go through to construct each different type of process map. Describing one all-purpose method for mapping processes is not our intention. Nevertheless, the overall mapping processes share many similarities. One of the purposes of this chapter is to describe these common traits. At the same time we will point out that there are many variations of this process and that adaptations must be made where necessary.

Although there are many dissimilarities in the different types of process maps, for example, in the information they contain and their actual construction, the "meta-process" is actually more important. First of all, a good process for mapping a process will contribute to ensuring a high-quality map that displays information that truly represents the process in question. Sometimes, however, the primary value of a mapping exercise is not the map itself, but what the map leads to in terms of consciousness and discussions that in turn often lead to improvements in how processes are carried out. In many of the instances where we have taken part in mapping exercises, mapping has been more about instigating change than it has been about providing a graphical account of the processes. Even in such cases, however, the map has been greatly appreciated as a tool to ease communication about elements of the process in question. Constructing a map puts words to and clarifies elements that might otherwise be wrongly assumed, and the visual nature of process maps allows pointing to activity flows or steps and discussing them.

# 3.1 THE META PROCESS

The overall mapping process (the meta process) can be divided into the following phases:

Plan

Execute

Implement

These phases are briefly explained below.

## Plan

This phase in described in the current chapter. Important elements are:

*Clarify the purpose*

- Clarify the rationale, purpose, and method
- Select the process
- Select the team

## Execute

This phase is mainly described in Chapters 4 through 8, covering the creation of the following process map types:

*Define the process(es)*

- Stakeholder map
- Value chain map/high-level process model
- Basic flowchart
- Cross-functional flowchart
- Bottleneck map/flowchart with load statistics

Please bear in mind that these different types of process maps could be performed sequentially. However, this is not always the case. Often, we only need to use one or two of the map types in an iterative process where the sequence might vary.

## Implement

We have earlier stated that this book does not focus on implementation; we should mention that an important subtask in the implementation phase is to create acceptance for the suggested changes and a favorable climate for their implementation. This is a significant task that involves disciplines such as psychology, human resource management, and so on. It is important to have a *rigorous* change approach. However, it is equally important to create acceptance of the change among those involved. We have seen many excellent mapping exercises fail due to lack of acceptance among those involved.

## 3.2 BEFORE YOU START CONSTRUCTING THE MAP

The first step in planning the mapping process is to define clearly why you would like to map a process. The rationale for mapping the process and what you are planning to use the process map for should influence how you plan and carry out the work of mapping it. Similarly, the approach will be different depending on whether you plan to map an existing process or a new process.

Based on what you define as the purpose of the mapping, you will have to define which process(es) you are going to map. The process(es) should be clearly defined, in terms of both departure point and end point, and what is included in the process. This might be challenging, as most processes interact with each other. In order to clarify this, it may be possible to regard the process as a work unit, and define the boundaries for this work unit. The input from suppliers to the process should be clearly defined, as well as the output with the requirements of the customers.

There is also the question of how many processes you will need to map for the stated purpose, and in which order they should be tackled. There is no easy way to determine this, but we have found that a simple prioritization matrix can be helpful. It is based on relating work processes to key business objectives, often termed critical success factors (CSF).

Typical examples of such critical success factors are the price asked for the organization's products or services, the quality of same, special features of the products or services, reputation, and so on. If asking the question, "What is it that our customers truly value about our enterprise and that helps maintain them as customers?" the answer will usually constitute the critical success factors.

A simple prioritization matrix lists the success factors on top, with candidate work processes that influence these in rows. The success factors can be assigned weights; in the example in Figure 3.1 these range from 1 (lowest priority) to 3 (highest priority). For each process, you ask the question, "To what extent does the process influence our ability to fulfill each success factor," again scored from 1 (lowest impact) to 3 (highest impact).

*Prioritization matrix for processes*

| Processes | CSF Weight | 1 3 | 2 1 | 3 1 | 4 3 | 5 2 | Total score |
|---|---|---|---|---|---|---|---|
| Process 1 | | 3 | 1 | 2 | 9 | 4 | 19 |
| Process 2 | | 9 | 3 | 1 | 3 | 2 | 18 |
| Process 3 | | 9 | 2 | 3 | 6 | 6 | (26) |
| . | | | | | | | |
| . | | | | | | | |
| . | | | | | | | |
| . | | | | | | | |
| . | | | | | | | |
| Process *n* | | 3 | 2 | 2 | 3 | 6 | 16 |

**Figure 3.1**   Prioritization matrix for work processes.

Multiply the impact factor by the weight assigned, place the score in the right cell, and summarize for each process. This numeric value indicates the collective impact of the process on the complete set of critical success factors. The higher the score, the better reason to map and improve the process, as this will give the greatest overall effect on the organization's critical success factors.

*A precise mandate makes it clear what to expect from the mapping exercise*

You, or anyone else acting as the principal, should aspire to provide a precise mandate so that everyone involved knows what to do and what to expect from the mapping exercise. This is particularly important if there are several persons involved in the mapping exercise.

Based on your specific reasons for mapping the process you should not only choose which type of process map to employ, you should also make choices regarding how to approach the construction of your map in terms of:

*Choosing your approach*

1. Who should be involved?

2. How should they be involved?

3. How much time can you spend on the task?

4. What are the relevant sources of information?

5. What level of detail should be used?

Each of these are discussed below:

1. *Who should be involved?* Most people having some knowledge of the process you are about to map will be able to contribute to constructing the map. In our experience, the task of mapping a process often requires a number of people with different types of knowledge and experience. Generally groups of five to eight persons are ideal as no one will "disappear in the crowd." When mapping processes, however, the requisite knowledge is more important than sticking to a predefined number of group members. If your group has more than eight members, it is important to make sure that everyone's opinion is being brought forward, for instance by having a facilitator to lead the group's work. Customers of the process, suppliers to the process, and the process owner should all be involved in one way or another. Making a stakeholder map, as shown in Chapter 4, is a good way to answer the question of who should be involved. Another question to consider is whether you need assistance in terms of an external facilitator.

*Who should be involved?*

2. *How should they be involved?* Involving a number of people can be done with group sessions, constructing the map from scratch, or in individual meetings or plenary sessions where individually prepared drafts are discussed. Sometimes simply drawing the map yourself and exposing it to peer review is fine. Our experience though, is that making the mapping task a group exercise has many tangible benefits. Group/plenary sessions can be held with or without facilitation (external or internal).

*How should they be involved?*

3. *How much time can you spend on the task?* This depends on the complexity of the process, the existing knowledge of the process, and the resources available. The latter implies how much time you and anyone else involved in constructing the map can set aside within the specified period of time. Additionally, it depends on when you need to complete the map. Lastly, it will also vary depending on whether the current process is new or not.

*How much time can you spend on the task?*

4. *What are the relevant sources of information?* Your process map can be based on various sources of information, such as documents, people, and observations. The relevant documents could include the quality assurance manual, work descriptions, procedures, manuals, production data, the information and communication system, and the management system. Please keep in mind that these all provide descriptions of how the process should be conducted, and they might not correlate fully with how it is really performed. In order to investigate how the process is really conducted, you can observe and interview people. These would then serve as sampling tests in order to validate the documented process. Often, a combination of observations, interviews, and informal talks with key personnel provides the best information.

*What are the relevant sources of information?*

*What level of detail should be used?*

5. *What level of detail should be used?* This all depends on the situation at hand. We do, however, advise that the level of detail should enable you to understand the process, not just scratch the surface. A common fault is to go into too much detail, and thus lose sight of what is really important. A middle course often works well.

## 3.3 KNOWING THE TEAM

*Get to know each other*

Once the team has been established, it is vital that the team members get to know each other. One way of doing this is to have a social aspect to the kickoff, or to meet one evening for social activities. We have often found it fruitful to meet at a local bar, or to get together and do such things as bowling, go-karting, or watching a sports game. This enables the team members to feel more comfortable in the group setting, which is vital for the interaction of the team. Once you know each other it is easier to communicate and to give constructive feedback. As a result, the group process will be more effective, and this approach also reduces the risk of conflicts within the team.

It might also be a good idea to keep a team roster, listing the names of the members, contact details, and their attendance records. We have provided an example of a template for this in Figure 3.2.

**Team Roster**

| Name (role) | Telephone | E-mail | Meeting dates and attendance | | | | | |
|---|---|---|---|---|---|---|---|---|
| 1. | | | | | | | | |
| 2. | | | | | | | | |
| 3. | | | | | | | | |
| 4. | | | | | | | | |
| 5. | | | | | | | | |
| 6. | | | | | | | | |
| 7. | | | | | | | | |
| 8. | | | | | | | | |
| 9. | | | | | | | | |
| 10. | | | | | | | | |

**Figure 3.2**  Team roster template.

## 3.4 GROUP ROLES AND FACILITATION

Do not organize the task in a more complex manner than is required. You can perfectly well construct a process map in a group without assigning different roles to the group members. If you do assign roles to members of the group, these are the roles we recommend that you consider including:

*Keep it simple*

- Facilitator, to lead the group process by clarifying the purpose of the meetings and describing roles, working methods, and scope. The facilitator is also responsible for making sure that the discussions stay relevant to the process in question and that they are aligned with the purpose of the meetings.

*Roles in group meetings*

- Secretary/recorder, to document the discussions and the resulting outputs.

- Group leader. Especially for mapping complex processes, it might be advisable to have a group leader (in addition to the facilitator). If so, it is important that the distribution of responsibility between the facilitator and the group leader is clarified. The group leader should preferably be a member of the organization directly related to the process being mapped.

- Team members, to provide knowledge of the process and participate in the discussions.

- Experts, who have deep knowledge of certain aspects of the process that can supplement the team's insight where needed.

The roles of facilitator and secretary are both quite demanding, so we do not recommend combining these roles with the role of ordinary group member. Again, all of these roles will not be required in all mapping tasks. We recommend, however, that you consider making use of a few specific roles in mapping situations where:

- Tension or conflict is expected among group members. Here, a facilitator, maybe even an external facilitator, may be needed to stop dead-end discussions and arguments that are not contributing to the work of the group.

- The level of detail in the discussions is expected to be higher than what you are able to include in the map. (Having a secretary will enable the group to take down important, detailed elements of the discussion that are not easy to implement in the graphical account of the process.)

- The time available for the mapping exercise is very limited; a designated facilitator with a clear plan for the group meetings can help the group comply with the mandate within the time limit.

- You plan to involve people outside the group between group meetings, for instance to comment on draft process maps or contribute input on issues concerning parts of the process. A group leader can be given the task of involving people outside the group in this way.

*Facilitation*

A key task for the facilitator is to make sure everyone gets a chance to participate. She or he should ask questions directly to people who are silent, and must not be afraid to reign in excessive talkers. The facilitator should always focus on the task at hand and remind the others to keep the overall goal in mind.

The group process will go through several phases. In cases where the group needs to be creative, tools for idea generation should be used:

- *Brainstorming* is the most basic of these techniques, where the focus is to generate as many ideas as possible.

- *Brainwriting,* also known as the *Crawford slip method,* is essentially a written version of brainstorming.

- *Nominal group technique* is a more formalized way of generating and voting on ideas.

- *Affinity charts* are used to organize thoughts or ideas.

For details on how to apply these tools, please consult relevant books from the recommended literature at the end of this book. Even if you have chosen not to formally assign the facilitator role to anyone, it can be helpful for the team to know some of these techniques so they may resort to them if discussions slow down too quickly. All of the tools have advantages and disadvantages; brainwriting is, for instance, a suitable tool if you have dominant people in the group.

*The parking lot*

Sometimes ideas that are not suitable for the discussion come up during the group meetings. The discussion may take a turn that moves it away from the topic or raises a question that is not suitable for the group (for example, beyond the group's scope or for which the group does not hold the required competence). Discontinue a discussion that does not serve the group's purpose and set it aside in "the parking lot." The parking lot, preferably a flip chart visible to the group, serves as a notepad for concepts that cannot be discussed further in the group at the moment. The parking lot is a powerful tool for the facilitator to allow the group to move away from a topic that for any reason is not suitable for further discussion.

For stopping people from talking too much, going beyond the scope, or in other ways obstructing the group's work, yellow and red cards may be useful. The idea is based on the cards soccer referees display when players violate the rules of the game, and the facilitator or group members indicate this by holding up a colored piece of paper. A yellow card is the milder signal, warning people that further breaches of civilized conduct will result in the red card. The red card sends the member "off the field," meaning banishment from the meeting for either a period of time or for the remainder of the meeting.

*Yellow and red cards*

## 3.5 MEETING LOCATION AND AIDS

As mentioned earlier, the facilitator should make sure that the group stays focused. A practical measure in this regard is to meet away from the office so that group members will not be caught up in checking their e-mails or engaging in discussions with colleagues. All too often we see meetings where the attendees have their PCs, BlackBerrys, or other devices turned on and are actually checking their e-mail and working on other projects during the meeting. This provides a poor working climate for the group. Lastly, phones should be turned off and the group should be encouraged to be back on time after breaks.

*Setting*

It is always an advantage to agree to principles on how to interact in the group, the "group operating principles." These rules could be discussed orally, written on a flip chart, or prepared in advance by the facilitator. Usually, we find it best to discuss the rules and to write them on a flip chart. The rules should include aspects like how we interact, how we listen, how we communicate, and how we decide. Additionally, rules regarding how to discuss problems should be dealt with.

*Ground rules*

There are a multitude of ways to organize a meeting room. Which solution is chosen depends on the type of discussion planned for the meeting. Typical meeting room designs include:

*The room*

- Chairs in a semicircle and no table. This focuses attention, but many people are uncomfortable without having a table in front of them. (See Figure 3.3a.)

- Traditional meeting room, with chairs around a table as people are used to. This is how most meeting rooms are arranged from the

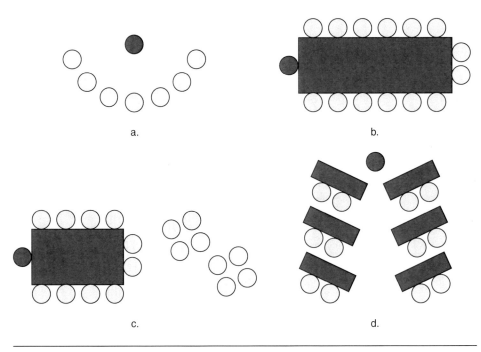

**Figure 3.3** Typical meeting room designs.

beginning. However, this setup is inflexible and it does not encourage discussion across the room if the room is large. (See Figure 3.3b.)

• Combined table setup with chairs for small group activities. This is a flexible setup that can be changed during the session. However, it requires a large room and is best suited for large groups. (See Figure 3.3c.)

• A V-shaped setup with tables in front. This is a good setup for medium-sized groups. It provides group members with a writing surface during the meeting, and it is easier to discuss across the tables. A disadvantage is that the design takes up much space. (See Figure 3.3d.)

The following equipment usually comes in handy in meetings:

*Materials*

• Markers of different colors.

• Flip chart.

• Tape. With masking tape you can move objects (processes, inputs, outputs) around your map several times. Use ordinary clear tape to fix the objects in place.

- A large roll of paper or a large piece of cardboard.

- Sticky notes, if possible in different shapes and colors.

- Pens.

- Beverages and fruit.

Some advice for the facilitator before starting the session:

- Arrive early.

- Check the seating arrangement and possibilities for reorganizing it.

- Test any audiovisual equipment and lights to see that they are in working order.

- Check the markers to see that they all work.

- Check that you have the necessary materials (see the list above).

- Make sure the flip charts have enough paper.

- Ensure that there is sufficient space for the planned activity.

*Tips for before the session*

## 3.6 ALL IN FAVOR?

What if group members disagree? Does the group have to reach consensus? Our simple answer to this question is no. However, there are different levels and meanings of disagreement, especially depending on whether you are describing an existing process or a future/theoretical process.

*Need to reach consensus?*

If people mapping *an existing process* describe the same part of it differently, this means that either someone has got it wrong or that the process is actually not performed in the same way on every occasion. Realizing that the process is not as streamlined as you may have thought is an interesting finding in itself. In this case, disagreeing is OK—it helps bring forward supplementary information.

When describing *a future process* there may be different opinions on what is the right way to perform the process (how to organize it, what technology to use, whether to outsource functions, and so on). In our experience it is important to set up the roles so that the group is not a decision-making body. When describing future processes, results from the group should be considered as advice to decision makers. Differing opinions as well as the reasoning behind them must, however, be made available to the decision makers through the documentation prepared by the secretary. Sometimes it will be necessary to agree to disagree so that the work can proceed.

# 3.7 ORGANIZING THE PROCESS MAPPING AT BROOK REGIONAL HOSPITAL

In the following example we have not included the process map resulting from the work described. This was done to maintain focus on the topic of this chapter—*the process* of mapping a process. The subsequent chapters of this book will attend to different types of process maps.

Brook Regional Hospital would like to prepare for the possible future construction of a new hospital wing accommodating pediatrics. Construction of a new wing had been discussed for some years in BRH, but the formal decision to build had not been taken. The board of directors stated that they wanted a rough, realistic picture of what the hospital would be faced with in case one day the decision would be made to build the new wing. They mainly wanted information that could serve the following needs:

1. Provide input to the architects at the initial phase of the design process.

2. Prepare for the organizational development initiatives that would inevitably follow once pediatrics had the opportunity to inhabit their new quarters.

Assistant Director Watts was assigned these tasks. Watts decided to map the main pediatric process with the help of some of his colleagues from that department. The completed process map would be useful, but he considered the content of the discussions with his colleagues to be equally important.

As building the new wing had not been sanctioned yet, the hospital did not want to put more resources into the mapping exercise then necessary. This meant that the group would identify the big picture and some key elements, rather than spending its time on details and finding solutions to every issue that could come up. It also meant that the group would include in-house resources only, and no external consultants. Watts himself chose to lead the group's work to make sure that the group would stay focused on the assigned tasks.

To ensure a group that would be able to see the processes of the pediatric practice from different perspectives, a relatively varied group was selected, consisting of the following people:

- The chief pediatrics physician

- A consulting pediatrics physician

- A head nurse

- An imaging operator

- A laboratory technician

- A pediatrics clerical assistant

For the time being, Watts considered himself and these six to hold the relevant knowledge. In a situation where the decision to build the new wing had actually been made, Watts would have included an architect and a user representative in the form of a parent of a former patient. Watts convened a group meeting, and opened the meeting by clarifying the group's mandate. He also emphasized that the group members were chosen because of their knowledge and experience relevant to the process, and that they were not representing departments or their formal personal roles in the organization. An experienced nurse was assigned the role of secretary to document the group's discussions and conclusions in writing. In this first group meeting, lasting for four hours, the group constructed a draft of the main pediatric process. This was done by attaching pieces of paper illustrating activities to a large piece of paper covering the wall of the meeting room. Adhesive notes containing supplementary information were added to this map. Different colors were used to mark categories of information: yellow notes for collaboration with other departments or units within the hospital (imaging, laboratory, specialized medical departments, and so on), green notes for activities the group regarded as good practice, and red notes marking potential improvements.

At the end of the first meeting, the group agreed that they should collect more information regarding some of the activities before the next meeting. More precisely, they needed more insight into the collaboration with another hospital on patients with a certain diagnosis, as well as how the orderlies influenced patient logistics in the main pediatric process. The group also decided that the group members would discuss the draft process map with some of their colleagues to get some feedback.

The second group meeting took place two weeks after the first one, and the process map was completed based on feedback and additional information from people outside the group. Ideas for improvements that could be introduced even in their present quarters were identified. Watts and the secretary had prepared for the second group meeting by sorting the information from the first meeting into two groups: one for input to the architects at the initial phase of the construction planning and one to help prepare for the organizational development initiatives. Thanks to this, the second group meeting lasted only two hours.

Watts and the secretary wrote a report based on the group's work and sent it to the group members for comments. Five weeks after Watts was given the task, he sent a revised report to the board of directors. In addition to the map of the main pediatric process, the report included what the group at this point considered preferred solutions for organizing the department. The group also identified a number of central requirements for the physical layout of the new wing. As they had gone through the process relatively quickly, the group had come up with more questions than answers. The unanswered questions were also documented in the report, so that these could be looked into more thoroughly should it be decided to realize the new wing. The board of directors was particularly pleased with a separate chapter in the report pointing out improvements in organizational practice that could be introduced within the existing building.

# Chapter 4

# Creating a Stakeholder Map

T he next five chapters will explain how the different types of process maps can be applied in a comprehensive manner, where each level of process map builds on the previous one. Our purpose is to show that modern process mapping is both more powerful and more complex than simply drawing a flowchart of the sequence of steps in a limited activity. The way these chapters unfold, using the Brook Regional Hospital as illustration, shows how higher-level mapping of the organization's place in its environment influences which work processes are mapped and how. Although it's stretching the concept a bit to label a stakeholder map as a type of process map, we hope you will appreciate how starting at this high level plays an important role in the overall process mapping exercise.

*Stakeholder*

## 4.1 WHAT IS A STAKEHOLDER MAP?

The term *stakeholder analysis* is probably better known than *stakeholder map*. Stakeholder analysis is by now well established as a technique for scanning the external and internal surroundings of the organization with the purpose of understanding which individuals, companies, institutions, and authorities our organization must relate to, and what expectations they hold of us, thus dictating their likely position in their dealings with us.

The stakeholder map is simply a graphical representation of the stakeholder analysis. In its simplest form, it depicts the organization and the stakeholders that surround it. A generic example is shown in Figure 4.1.

If you have never given the organization's stakeholders any thought before, such a simple map might provide some new insight. For most organizations, however, we can assume that the information given in this map is

29

*Simplified, generic
stakeholder map*

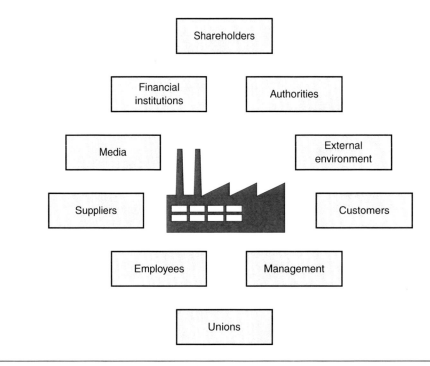

**Figure 4.1**  Simple, generic example of a stakeholder map.

already well known. In our experience, there are many eye-openers to be discovered when discussing the expectations held by the various stakeholders.

*Stakeholder
expectations can
hold surprises*

We might assume that suppliers, being grateful for the business we generate, are always pleased with us and hold no other expectations than continued future business. With many suppliers becoming as discerning as customers in who they choose to do business with, you might find upon closer examination that some of your suppliers are considering focusing their position in a couple of supply chains, thus declining your offer of a prolonged relationship. Or, closer inspection of customer expectations could reveal preferences to date unsatisfied by your products or services, not to mention employees (internal customers) no longer content only with being paid for their efforts, but rather expecting active lifelong learning mechanisms and clear career paths.

Such information can easily be added to a stakeholder map, as the example in Figure 4.2 shows.

An additional element often included in a stakeholder analysis is an assessment of the position or attitude of the stakeholders toward us. There are many models available that outline different ways of classifying stakeholder positions, but here we will settle for something as simple as:

*Stakeholder positions*

• Positive, interested in cooperation for mutual benefit

- Neutral, no strong position or attitude toward cooperation

- Negative, sees benefit for itself in disadvantage for us

Adding this type of information to the stakeholder map by using color codes makes it even more illustrative and useful as a tool in strategic planning, self-assessment, and so on. An example of this added functionality is shown in Figure 4.3.

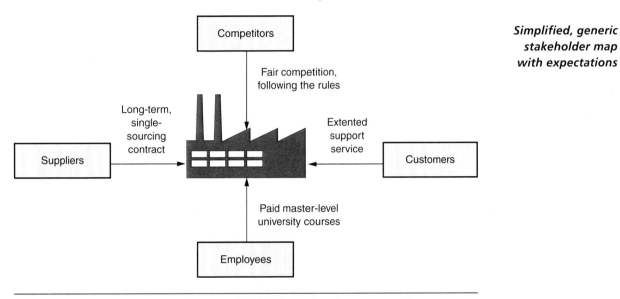

*Simplified, generic stakeholder map with expectations*

**Figure 4.2** Stakeholder map supplemented with information about expectations.

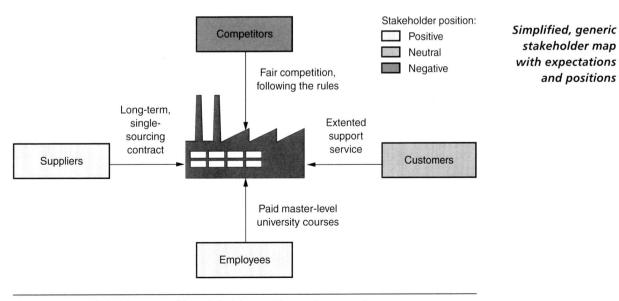

*Simplified, generic stakeholder map with expectations and positions*

**Figure 4.3** Stakeholder map with stakeholder position and expectations.

# 4.2 CONSTRUCTING A STAKEHOLDER MAP

A full stakeholder analysis, utilizing more complex approaches to assessing stakeholder needs and attitudes, is a rather complicated exercise. Such an extensive analysis is often included in strategic planning applications, but for our purpose—using stakeholder analysis as a preamble to more detailed work process mapping—a simplified approach will suffice. Thus, the steps in constructing a stakeholder map are:

*The steps in constructing a stakeholder map*

1. Identify the organization's various external and internal stakeholders, typically through a brainstorming session. On the stakeholder map, place your organization at the center and place the identified stakeholders around it. If possible, place the stakeholders in a logical pattern on the map, for example, suppliers on the left-hand side, customers on the right, "superiors" like owners, authorities, and so forth, above, internal stakeholders grouped together, and so on. Use arrows to visually link the stakeholders to your organization.

2. Analyze the needs and expectations held by the stakeholders, again either by brainstorming or in some cases by actually asking them directly. Include at least the most important expectations of each stakeholder in the map by listing them along the arrows linking them with your organization.

3. Assess the position of the stakeholders toward your organization. Color the different stakeholders on the map according to the identified positions, using green for positive, yellow for neutral, and red for negative.

This procedure will be illustrated in the BRH case shortly.

*Practical advice*

First, some practical advice and hints about constructing a stakeholder map:

- Two common mistakes in stakeholder mapping are to include too few or too many stakeholders. Too few means important stakeholders that truly influence the future of the organization can be overlooked, too many can make the stakeholder map unduly complex and hide, rather than reveal, important insight.

- If you think too few stakeholders have been identified, try to include other people in the exercise or systematically review categories like public authorities, financing institutions, customers, suppliers, R&D partners, educational institutions, nonprofit organizations/ nongovernmental organizations, local community, media, owner, employees, and so on, to identify further actors. Remember that important stakeholders can sometimes be "silent" and easy to overlook.

- If you find the stakeholder map getting to be cluttered with too many players, try to eliminate some by asking which of the stakeholders are critical or in a position to influence the organization and its future significantly in various situations. Also remember that several seemingly less important stakeholders with similar or identical expectations toward your organization can carry as much weight as one obviously central player.

- Don't be afraid to consult stakeholders if you are uncertain about their position and future plans. A stakeholder map founded on mistaken assumptions about stakeholders can be a dangerous starting point for your future work on work processes, and in most cases, stakeholders appreciate being consulted.

- There are more advanced approaches to stakeholder analysis, including ways of assessing stakeholder needs. If you would like to explore these further, please see the recommended literature list for relevant texts.

## 4.3 MAPPING STAKEHOLDERS FOR BROOK REGIONAL HOSPITAL

Following the announcement that People Care Hospitals will take over BRH, the hospital management realizes that this will introduce major changes. Having some knowledge of the PCH group and how they run their hospitals, it is clear that new requirements will be imposed and that the days of being a "sheltered" local hospital are over. Instead of remaining passive and simply accepting what the buyer dictates, however, the management team decides they want to more actively influence the situation by positioning the hospital for its future role.

It is decided to start the positioning process by performing a quick stakeholder analysis, mainly to understand what forces surround the hospital. A secondary motivation is to start a process of changing the mind-set of a predominantly introverted management team to a more active, extroverted focus.

*BRH stakeholder analysis*

The managing director summons her extended management team, consisting of two assistant directors, the head of administration, operations, and maintenance, as well as the unit managers for the different care units. They start the exercise by brainstorming stakeholders and end up with the primary stakeholders presented in Table 4.1 (the list was longer, but has been abbreviated for clarity). Moving on to identifying expectations, key elements for each of the stakeholders are also listed in the table (again, many more elements were identified, but these have been omitted). Finally, an assessment of each stakeholder's position is indicated in the final column of the table.

The resulting stakeholder map that emerged from this exercise is shown in Figure 4.4.

**Table 4.1** BRH stakeholders with expectations and positions.

| Stakeholders | Expectations | Position |
|---|---|---|
| People Care Hospitals | • BRH finds it place in the group<br>• Turns a profit<br>• Not exposed to any scandals | Positive, but demanding |
| Other hospitals in the PCH chain | • BRH finds it place in the group<br>• BRH does not threaten the existence of other hospitals | Negative |
| Competing hospitals outside the PCH chain | • Expects BRH to respect "division of labor" in terms of sharing patient groups | Neutral or negative where potential competition is seen |
| Primary care physicians | • BRH takes referrals seriously | Positive, but concerned |
| The local community (county, town, inhabitants) | • BRH remains a strong community healthcare provider | Positive, but concerned |
| Different patient groups | • To be served locally and not having to travel far to other hospitals | Positive, but concerned |
| Healthcare authorities | • BRH follows accepted standards and maintains quality care | Neutral |
| Suppliers of products and services | • Continued business<br>• Acceptable prices and terms | Positive, but concerned |
| Local media (newspapers, radio, TV, and so on) | • Access to news about BRH | Neutral |
| Ambulance services in the area | • BRH handles emergency cases as before | Positive, but concerned |
| Public social service administrations | • Close cooperation with BRH in cases where this is necessary | Positive, but concerned |
| Different groups of hospital employees | • Secure jobs<br>• Equal or improved working conditions as before | Positive, but concerned |
| Volunteers active within the hospital | • Equal or improved working conditions as before | Positive, but concerned |

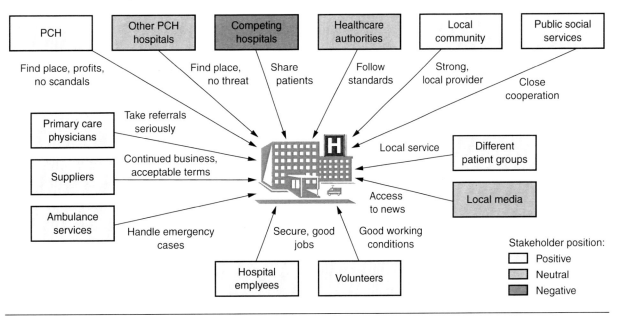

**Figure 4.4** Stakeholder map for Brook Regional Hospital.

Having completed the stakeholder mapping, several issues became much clearer, including the following:

- Many stakeholders felt strongly that the hospital should stay a local hospital catering to the community's needs; an equal number of stakeholders were concerned that things would change for the worse.

- The PCH group, on the other hand, expected BRH to both improve its efficiency and find a niche within which it could become a leading provider. The flip side of this insight was that staying a very broadly oriented hospital, serving large numbers of different patient groups, was a bleak prospect with PCH as owners.

- Other hospitals, both inside and outside the PCH group, who had lived with more open competition for years, felt threatened that BRH might possibly squeeze them out of their niches.

In light of this new insight, the hospital director invited the PCH management team to BRH for a tour and discussions about the future. The visit confirmed their demand that BRH identify niches where it could compete successfully, but also revealed that high performance would open up opportunities for expansion, even the construction of the new wing, which had been discussed for at least ten years within BRH. Pediatrics had been targeted by BRH for upgrading, but the unit had lived with poor facilities for many years.

## Post-Case Reflection

Seeing how BRH used the stakeholder map, we hope you realize that such a map is quite powerful in allowing the organization to obtain a clearer view of itself and its surroundings. From this improved vantage point, the stakeholder map can also be used as part of strategic planning. And a typical side effect of this mapping process is that the organization, or at least parts of it, realizes that change is required to remain viable in the future. This helps motivate both the analytic effort required to devise improvements and their subsequent implementation, another side effect whose value should not be underestimated.

# 4.4 CHECKLIST

*Checklist for stakeholder mapping*

❑ You have composed a sufficiently broad team to undertake the construction of the stakeholder map

❑ You have brainstormed stakeholders in different categories

❑ If too many stakeholders were identified to handle properly in the map, you have eliminated the least important ones

❑ You have understood the position and attitude of the stakeholders sufficiently, possibly through consultation

❑ You have discussed the assumed position—positive, neutral, or negative—of the stakeholders

❑ The stakeholder map depicts the actual stakeholders, their most important expectations, and their position toward your organization

❑ Based on the stakeholder map, you have discussed the implications for your organization and which work processes must perform well

# Chapter 5

# Creating a Value Chain Map/High-Level Process Model

In the progression through different levels of process mapping, the stakeholder map leads logically to a more focused *value chain map,* supplemented by a *high-level business process model.* Let us, however, again emphasize that there is no law dictating that you need to proceed through all these different levels; depending on how much mapping you have done previously and the purpose of the mapping, you can skip directly to one of the more advanced map types, or select only a few that you think will serve the intended purpose.

## 5.1 WHAT IS A VALUE CHAIN MAP/ HIGH-LEVEL PROCESS MODEL?

These are two different but intrinsically linked types of high-level maps, and they will be described one at a time. Very briefly, a *value chain* is a chain of organizations linked in supplier–customer relationships that join forces to deliver a service or product to an end customer. Value chains have been given various different names, including supply chains, demand chains, and so on. We will use the term value chain in this book, meaning a chain like the example shown in Figure 5.1, where the publishing house is the focal point of the map. Please notice that this is a very simplified example, where there is no indication of where the word processing equipment, paper, ink, truck, or bookstore furnishings come from. In fact, in most cases it is far more correct to talk about a *value network* than a value chain.

At any rate, the purpose of such a map is to illustrate and understand better where your organization fits into the larger picture. In this example, it is obvious to everyone in the value chain that they depend on each other and

*Explanation of value chain*

*Value network*

*End customers of the value chain also dictate your revenues*

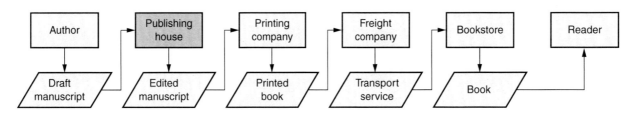

**Figure 5.1**    Example of a value chain for a publishing company.

are linked to one another, directly or indirectly. In many other situations, this is not always as easily apparent. If you are a company making sheet aluminum, you naturally know who your immediate customers are: typically companies that either produce finished products from the sheets, for example, soda cans, or that produce intermediate components of different types. Some might make housings for products like staplers, cellular phones, or DVD players; others use the sheet aluminum only as a small component in a more complex product, for example, a battery. The batteries are in turn sold to many different companies who use them in a wide range of products. Ultimately, knowing where your aluminum ends up is quite difficult but still important to know, as the market development for the end products dictates the sales of your aluminum.

A value chain map can be constructed in many different ways; in contrast to basic flowcharts, there is no single recommended format for it. The essential message it must be able to convey is:

*Elements to cover in a value chain map*

- The structure of the value chain, that is, the different organizations that comprise it and how they are positioned with regard to one another.

- Your organization's position in the value chain.

- The outputs (products and services) that are delivered from one link in the chain to the next.

Figure 5.1 showed one way of presenting this information; another example is shown in Figure 5.2, where the pet food producer is the focal point of the value chain.

What you will notice from both these examples of value chain maps is that they are quite rough, presenting only the different value chain members and their outputs. For the most part, this is quite acceptable, even beneficial, for giving a "helicopter" view of the flow of products and services through the chain. Especially for your own company, however, this is too high-level to provide much insight about how you manage your position and responsibilities within this structure.

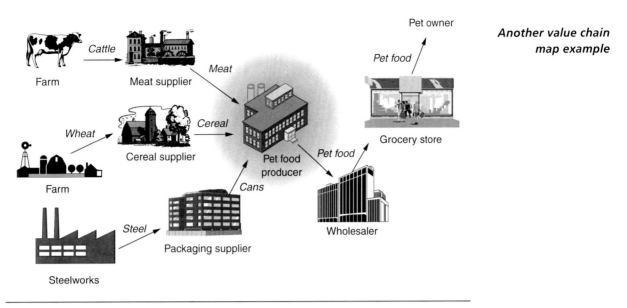

**Figure 5.2** Value chain for pet food producer.

If you "zoom in" on the pet food company in Figure 5.2, you will find a miniature value chain inside the company. This internal value chain consists of a set of primary work processes that deliver various outputs to the next process in line, as well as processes that support the primary, value-adding processes. Many organizations struggling to maintain high performance levels or even survive at all can attribute their problems to a mismatch between the stakeholder expectations they are faced with, their position in the overall value chain, and their internal value chain of work processes. Before going into the details inside each work process, it is therefore essential that you understand which primary and support processes you need, what these should deliver in terms of output, and how they must interface with one another.

*The internal value chain of the organization*

Digression: When Michael E. Porter coined the term *value chain* in the early 1980s, he referred to the internal chain of work processes that deliver value to the immediate customer of the company, whereas the extended structure of several organizations cooperating to deliver value to the end customer was termed *value system*. The usage of the terms has gradually changed; today most references to value chain mean the extended supply chain.

*Michael E. Porter's definition of value chain*

This overall work process logic is most suitably depicted in a so-called *high-level process model*. In terms of value chain maps, such models can be constructed in many different ways. One approach is to draw it the same way as

*Explanation of high-level process models*

*Another value chain map example*

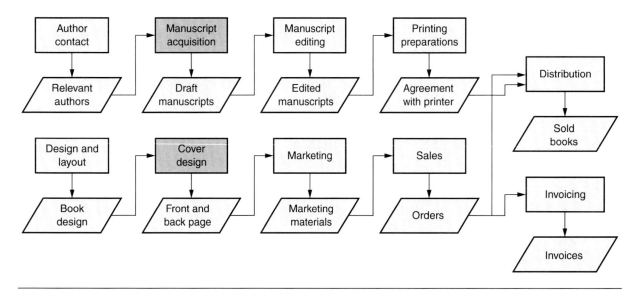

**Figure 5.3**   Example of a high-level process model for a publishing company.

the value chain map, as Figure 5.3 illustrates. In this case, a simplified view of the internal work processes of the publishing house from Figure 5.1 has been added in (notice that these are more or less only primary processes, no support processes have been included).

Being a rather simple example, this process model might not give rise to any breakthrough revelations about the suitability of the work process design of the company, but it still communicates a high-level view of the organization and allows people in different work processes to see their role in the larger picture.

An alternative approach to drawing the high-level process model separates the different types of processes. An example is shown in Figure 5.4 for the pet food manufacturer from the value chain depicted in Figure 5.2. Being again a simplified model compared with a real company, it excludes quite a number of processes, but shows how processes under the categories of value-adding, support, and management are linked to form an internal value chain that will deliver products to the customers.

## 5.2 CONSTRUCTING A VALUE CHAIN MAP/ HIGH-LEVEL PROCESS MODEL

These two types of process maps are different entities, but they naturally reinforce each other. We therefore normally recommend doing the exercise of constructing both at the same time. This way, you ensure an understanding

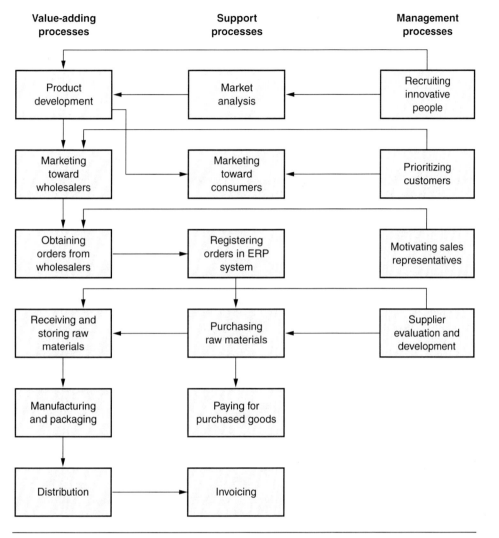

*Example of a high-level process model where different types of processes are grouped together*

**Figure 5.4** High-level process model for pet food company with process types grouped together.

within the organization of both the extended, external value chain and how this is supported by the internal chain of processes. We have therefore combined the steps into one sequence aimed at designing both models (but this is not to say it's impossible to do just one of them; if so, simply perform the required steps):

1. Clarify and list the organization's main products and/or services, and assess whether you are part of more than one value chain. If this is the case, decide whether to construct value chain maps for all or just one of them (if the former, repeat steps 2 through *X* as necessary).

*The steps in constructing a value chain map/high-level process model*

2. Place your organization in the middle of a blank value chain map along with the products and/or services you deliver to your customers in the value chain.

3. For the raw materials, components, services, and so on, that you source from suppliers, work backwards to form the chains of organizations and outputs that in the end culminate in deliveries to your organization. In some cases, this involves placing only one link in front of you, in others there could be literally dozens. You must decide how far back it is worthwhile to go. Typically, for some inputs it makes sense to trace back through a number of links, for other more marginal inputs there is no point in going beyond your immediate suppliers.

4. For your products and/or services, perform a similar exercise, tracing forward through to your customers' customers, as far as it makes sense, to form links of outputs and customers. On this end of the value chain, it usually is worthwhile going all the way through to the original equipment manufacturer (OEM) for product value chains or the end service provider for service value chains. In cases where your outputs are marginal components in the deliveries to the end customers, this is perhaps not equally important.

5. Having completed the value chain map, zoom in on your own organization and decide whether to design a swim-lane type of high-level process model or a less stringently organized model. The ensuing steps will be roughly the same in any case, with the exception of the swim-lanes themselves.

6. Based on your stakeholders and your position in the overall value chain, brainstorm the value-adding internal value chain of primary processes that run from customer orders being received to deliveries being made, and draw these in a logical sequence according to the type of model you have decided to make. Indicate clearly, either with arrows or separate boxes, the outputs from each process.

7. Discuss which support and management processes are required to keep the value-adding processes running at the expected level of performance. Place these in the high-level process model and link them to the relevant processes that they interface with. Again, describe the outputs generated by the processes.

This procedure will be illustrated in the BRH case shortly.

First, some practical advice and hints for designing value chain maps and high-level process models:

- Many organizations, when starting to design a value chain map, quickly realize that the complexity of their external relationships is greater than they thought. Very often they are members of several value chains, and what they thought of as value *chains* turn out to be complex value *networks*. This is quite normal, and the trick is partly to differentiate between the separate value chains by constructing several maps. Furthermore, accept that the map will be a network map, but reduce the complexity where possible by not tracing too far backward or forward for marginal inputs or outputs.

- If you find it difficult to trace in either direction, beyond your immediate suppliers or customers, this is a clear sign that you have been too concerned about your immediate surroundings and have not understood that your competitiveness depends on the success of the entire value chain. In such a case, pursue an active initiative to obtain a larger perspective, usually by talking to your suppliers and customers about their links in the value chain. If necessary, you might have to undertake a brief study of your industry to understand who your end customers are and how you and your products and/or services relate to them.

- Remember that although organizations are linked together in supply chains or supply networks, it does not necessarily mean they always work together altruistically for the end customer. There will always be opportunism, hidden agendas, attempts at suboptimization, and so on, and a supply chain map can also be used to identify such.

- When designing the high-level process model, it is always easier to work from demand backward, that is, starting with the products and/or services your customers buy from you. Follow the trail backward through how they were delivered to the customer, the steps to produce them, obtaining the inputs needed for production, and so on. The same goes for support and management processes; identify them by the demand for them in other processes, not by brainstorming the processes you know you operate.

- Make sure that the processes you include in the high-level process model are real processes, not just small tasks. The distinction can be difficult to define, but processes are often characterized by involving several people and/or departments, consisting of several steps, and having a clearly defined output. Using too limited tasks in the model will make it too complex to allow for understanding the overall "business logic" of the organization.

- If the high-level process model, having been constructed from demand, fails to include all the different work processes you currently run, use this as an opportunity to investigate further. Often, processes, especially support processes, remain operational more out of tradition than an actual need for them.

## 5.3 UNDERSTANDING THE VALUE CHAIN AND HIGH-LEVEL PROCESS MODEL OF BROOK REGIONAL HOSPITAL

A couple of weeks after having hosted a visit from the PCH management team, the director of Brook Regional Hospital received an e-mail asking BRH to clarify their view on where their future position will be in the People Care Hospitals group, especially in terms of value chains spanning different, more specialized hospitals and other healthcare services that interface with BRH.

The hospital director informs the management team about the request, and the immediate response is negative. BRH has traditionally allowed the different departments to nurture strong unit identities, and they have normally catered to different groups of patients that only to a limited extent required interventions from more than one department. As a result, there has been little talk about cross-departmental patient flows, and the general consensus is that the stakeholder map they already constructed should be enough.

A follow-up phone call from PCH politely reminds the BRH director that this was not a suggestion, but rather an order. The BRH management team is summoned again, and they decide to go through with a perfunctory value chain mapping to silence PCH.

They realize that the different areas of care could represent different value chains, and start the exercise with a test-run using diabetes patients. Attempting to put themselves in a future situation where they have been forced to think about division of labor among different, specialized hospitals, as opposed to doing everything in-house like today, they end up with the value chain map shown in Figure 5.5.

The exercise was useful in making the BRH management team appreciate both how complex a path the patient has to travel, sometimes both before and after entering BRH, and that there could be benefits in focusing even

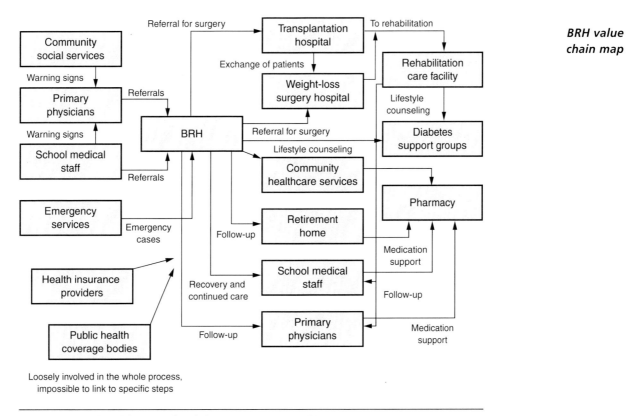

**Figure 5.5**   Value chain map for diabetes treatment at BRH.

more in depth on the prevention of diabetes and the facilitation of quality of life for diabetes sufferers. This would mean letting other hospitals assume responsibility for difficult and time-consuming surgeries that the BRH medical staff were not very comfortable with anyway.

Having come this far in changing their mind-set, BRH also admits that attention to the overall patient flow through the healthcare system could provide better care than having departments competing for resources, time, and the best working conditions. Although reluctant to succumb to every wish of the PCH management, they agree to review the high-level patient flow through the hospital, following up the diabetes treatment value chain map, starting with this patient group.

As they prepare for the task, the hospital director by coincidence learns that one of the anesthesia doctors, before entering medical school, had obtained a bachelor's degree in logistics. Having worked a couple of years as a logistics consultant, he was well trained in process flow mapping. Agreeing to support the management team in their mapping efforts, they proceeded to work together to produce an overall process model for BRH diabetes treatment, shown in Figure 5.6.

The process model quite clearly underlined that diabetes patients, who many prior to the exercise assumed were easy to handle and stayed mainly

**Elective diabetes treatment**

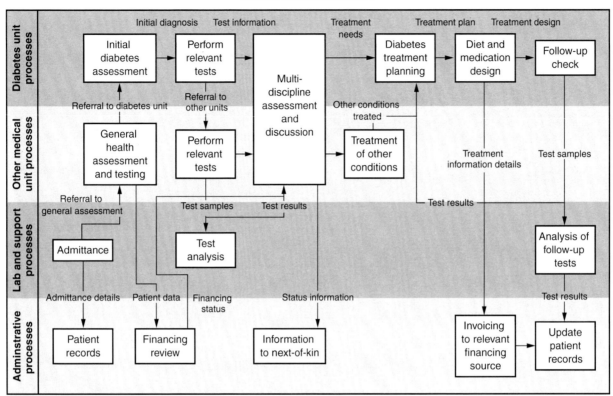

**Figure 5.6**   High-level process model for diabetes treatment in BRH.

within the diabetes unit, followed a much more complex flow through the hospital. Inputs were needed from various other medical units, especially the lab but also the administration unit. In this picture, they had even decided not to include support services like catering, laundry, maintenance, and so on, but recognized that these also played important roles in the efficient and effective treatment of diabetes patients.

At this stage, as the attitude toward the work imposed by PCH was gradually turning more positive, further insights were gained:

- Focusing the treatment on diagnoses especially suitable for BRH, patient groups or types of treatment that the hospital was not well equipped to handle could be transferred to more suitable hospitals.

- The interdepartmental interfaces and required cooperation were more complicated than previously believed.

- Some concerns were raised that patients frequently expressed dissatisfaction over being sent and back and forth, test results not being ready when a consultation was scheduled, missing paperwork, and so on.

## Post-Case Reflection

We hope you can appreciate how the value chain mapping exercise forces the organization to open its eyes and focus its perspective externally for a moment, thus realizing that it isn't an island that can live life as it pleases, but rather must find its place in a larger delivery structure. The high-level process model showed how assumed autonomous treatment areas actually are connected to other units in the organization, and that planning for a good flow through the whole process can be more important than suboptimizing within each area. Ultimately, this can even lead to reorganization away from strong functional units and toward more process-oriented teams/units that are set up to handle holistic work processes.

# 5.4 CHECKLIST

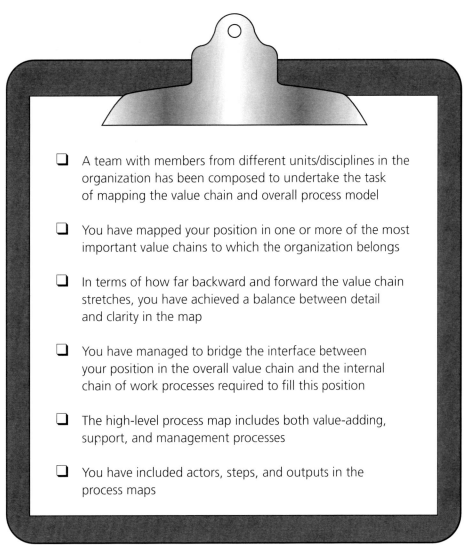

*Checklist for value chain mapping/ creating a high-level process model*

❑ A team with members from different units/disciplines in the organization has been composed to undertake the task of mapping the value chain and overall process model

❑ You have mapped your position in one or more of the most important value chains to which the organization belongs

❑ In terms of how far backward and forward the value chain stretches, you have achieved a balance between detail and clarity in the map

❑ You have managed to bridge the interface between your position in the overall value chain and the internal chain of work processes required to fill this position

❑ The high-level process map includes both value-adding, support, and management processes

❑ You have included actors, steps, and outputs in the process maps

# Chapter 6

---

# Creating a Basic Flowchart

The two preceding chapters placed our organization in a broader context by looking at external as well as internal stakeholders, and by focusing on value chains. Regarding our organization as one piece of a larger whole can be very useful. What we can actually influence, however, is often only inside of our own organization. Even though it can be used to describe processes along a value chain, a basic flowchart is more commonly used within an organization, or even within specific parts of that organization. As we indicated in Chapter 2, the basic flowchart is among the preferred types of process maps for many of the application areas we have outlined. The basic flowchart, as described in this chapter, can be a good starting point for mapping a process. Other mapping techniques presented in this book can help you elaborate on what you have visualized through the basic flowchart.

## 6.1 WHAT IS A BASIC FLOWCHART?

The basic flowchart is a simple map depicting the relationship between activities in an individual process. Like the other process maps illustrated in this book, it is a schematic representation of a process. Flowcharting is commonly found in business presentations. It is also a popular tool for people working with quality management. A number of software tools purposely made for drawing process maps are available. Today, you will even find flowchart functions in everyday software such as office application suites. These contain a variety of different symbols having a distinct meaning, depicting, for instance, processes, documents, decision points, data, manual operations, and—in our computer-based society—magnetic disks. Some examples of these symbols are shown in Figure 6.1.

If done in a consequent manner, using these symbols will surely contribute to a map that is rich in data. However, relying on more than a few symbols is not necessarily a good way to communicate with anyone who is not already familiar with the symbols.

In this chapter, we will describe a simple flowchart visualizing a process in order to make it more accessible. Rather than beginning with a wide range of symbols, we will in subsequent chapters extend the basic flowchart in ways that we have found fruitful in a number of situations. This will be done in the next two chapters, where we look at a cross-functional flowchart and a bottleneck map/flowchart with load statistics.

*Key*

Having decided to limit our use of symbols, we here introduce a rectangle describing an activity, a diamond describing a decision, arrows indicating outputs and inputs, and a square with rounded corners illustrating the start and end points of the process (see Figure 6.2).

The example shown in Figure 6.3 is of a general customer complaint process that could be found in many different industries.

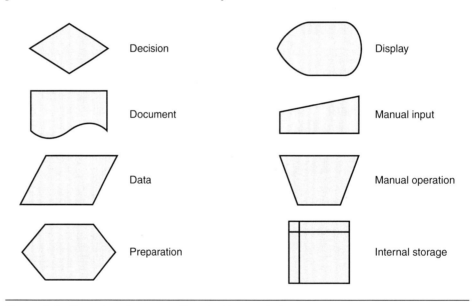

**Figure 6.1**   Examples of common flowcharting symbols.

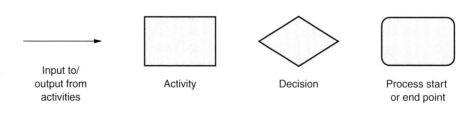

**Figure 6.2**   Basic flowcharting symbols.

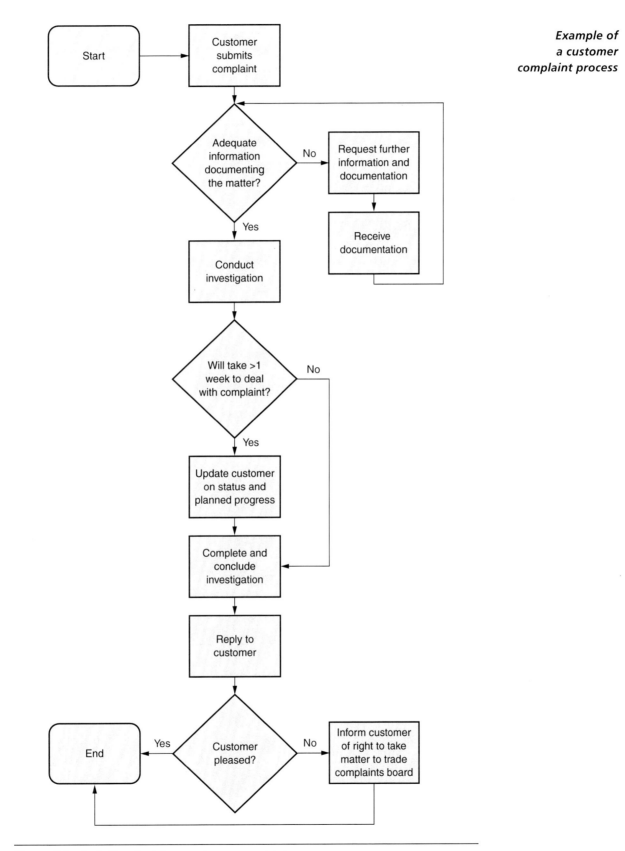

**Figure 6.3**  Basic flowchart for a customer complaint process.

# 6.2 CONSTRUCTING A BASIC FLOWCHART

Following are the main steps in constructing a basic flowchart:

*Steps in constructing
a basic flowchart*

*Direction of
mapping*

1. *Define the start and end points of the process.* Construction of a basic flowchart is based on the principles of defining the boundaries of the area to be analyzed and to start at the end of the process and work your way toward the beginning of the process. Working in this direction is particularly useful if you are constructing a new process or making a thorough revision of an existing process. When drawing an existing process or making smaller changes to a well-known process, it can be easier to start from the beginning of the process, working your way toward the end point.

2. *Select a graphical presentation format that suits your purpose.* Basic flowcharts can be formatted in many different ways; they do not have to look like the example in Figure 6.3. The main idea behind a basic flowchart is to make it as simple as you can as long as the map is constructed in a way that supports your purpose. In this chapter two different ways of presenting the same process within Brook Regional Hospital are demonstrated. If a more complex map is needed, you should use one of the approaches presented in subsequent chapters.

3. *Define boundaries between parallel processes.* When constructing a basic flowchart, it is important not only to clarify the start and end points of the process. As processes are often intertwined with other processes, you must also make sure that boundaries with other processes are clearly expressed. These boundaries do not necessarily have to be known from the beginning. When you start constructing the map, however, you will normally find rather quickly that you need to limit your focus.

4. *Break the map down into units you can handle.* Even basic flowcharts can become quite extensive. If you have a process that includes a large number of activities, it can be useful to break the process down into smaller parts while you are constructing the map. For instance, an activity comprising a large number of sub-activities can be treated separately before it is put together with the rest of the activities forming the total process.

5. It can also be useful to employ a template of questions that need answering, discussion, or clarification, either before the mapping is initiated or during the mapping exercise. An example of such a template is shown in the appendix. As you will see, this template covers many different aspects, from stakeholders to outputs to customers. Some of these are relevant for all types of process maps; others apply mainly to one or a few of them. When using the template, keep this in mind and select the elements that fit.

Before we illustrate how a basic flowchart can be formatted we would like to sum up some practical advice in designing basic flowcharts:

- Start your mapping exercise by a drawing a rough map focusing on the main elements of the process, adding details later. This helps you keep the process at a level you can handle.

- Do not add details to a basic flowchart until the flow depicted in the rough map has been agreed on.

- Drawing your map using software is fine for storing and distributing your map. If you draw the map in a group, our experience is that using software leaves too much influence with the person operating the keyboard. Furthermore, too much attention is often paid to software issues, such as font size and so on, instead of the mapping.

- Choose a map that suits your purpose—devise your own kind of basic flowchart if you find it practical.

## 6.3 CONSTRUCTING A BASIC FLOWCHART FOR BROOK REGIONAL HOSPITAL

Let us return to Brook Regional Hospital to see how they can employ a basic flowchart. The hospital has now been awarded the responsibility for several types of patient groups inside the larger PCH chain. The hospital chain has a tradition of measuring performance and imposing performance targets on the different hospitals. For BRH, they set targets for costs, treatment times, and quality.

The hospital, based on existing data, realizes that they have a ways to go before reaching the targets. A carrot to motivate improvements is that PCH has indicated that if the hospital demonstrates good enough performance, the chain will continue to develop the hospital as a specialized unit for the designated patient groups, and also will consider expanding the hospital with the desired new pediatrics wing.

Headed by the "logistics doctor," a team is assembled comprising doctors, nurses, lab personnel, administrative staff, and so on. To get an overview of their processes, they start by drawing a basic flowchart for holistic patient treatment. They define the starting point of the process to be when a patient is referred to BRH. The end point of the process is when patient treatment is completed. At the moment, this is a holistic process, and boundaries between parallel processes have not been defined.

Starting the mapping at the end point triggers a discussion within the group about defining what the term "treatment completed" means. The group finally agrees that not only the main treatment, but also any kind of medical follow-up after the main treatment, should be included in the process. This is a typical example of what happens in group settings: everyone had thought that the end point of the process (that is, patient treatment completed) was agreed upon, but discussing the process more thoroughly proved the need to be more precise in defining what this really means. The first version of the group's map is shown in Figure 6.4.

This first model is rather rough, and the group starts pointing out different routes of flow depending on factors such as whether or not the treatment is planned (for example, treatment for patients brought to the hospital from the scene of an accident can't be planned in advance). A decision or the answer to a specific question determines the next activity or decision in the process, as shown in Figure 6.5.

Even if this basic process map is on a high level (more detailed maps showing several steps within each activity surely could have been drawn), it reveals that most departments or units tend to suboptimize within their "four walls." One especially serious problem is found in relation to the activities

*Basic flowchart
focusing on activities
and process flow*

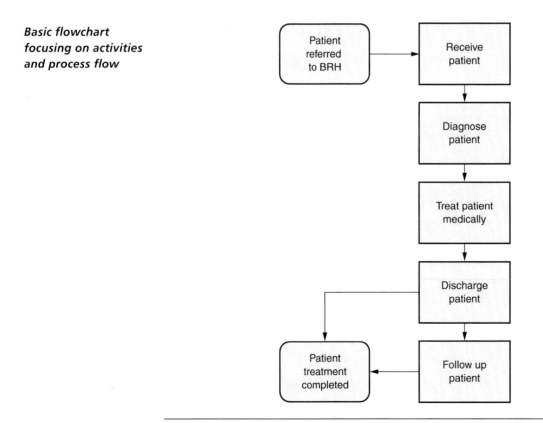

**Figure 6.4** BRH holistic patient treatment process—rough version.

of requesting and receiving further patient information and documentation. The discussion reveals that getting laboratory reports, which are one of the outputs of the more detailed activities under requesting/receiving further

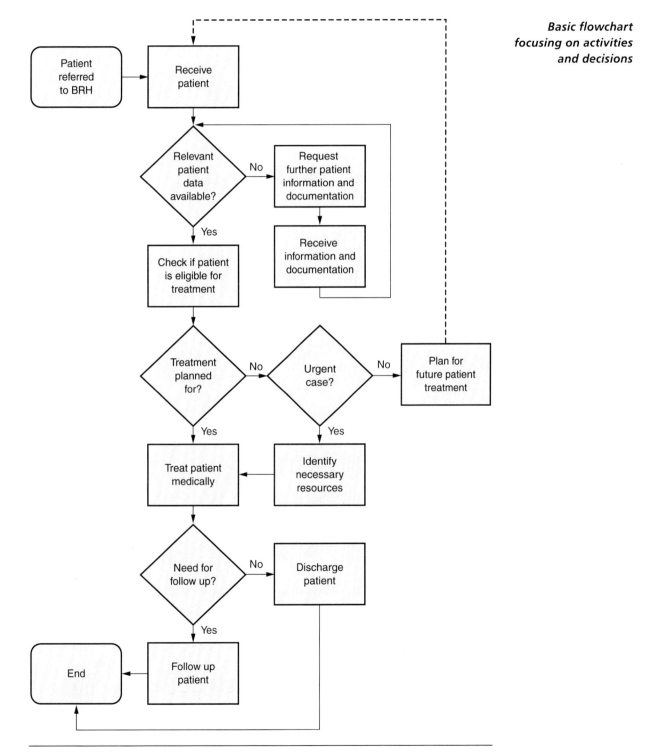

*Basic flowchart focusing on activities and decisions*

**Figure 6.5** Holistic patient treatment process—extended version.

patient information, can take an unreasonably long time. The laboratory is measured on costs per test performed, and tries to minimize these costs by assembling batches of identical or similar tests to gain scale advantages and minimize changeover times. This saves time for the lab, but without the lab understanding the consequences for other units—among others, surgeons having to wait for test results before being able to proceed with operations. There are also cases where the lab or other units or parties (for example, pre-op, psychologists) deem patients unsuitable for scheduled procedures but fail to notify other units about this since they don't see the consequences. In summary, the laboratory clearly did not understand its role in the overall process.

Another issue is a lack of standardization in procedures, which influences quality. In cases where problems occur, each person must improvise to find solutions, whereas studies from other hospitals show that predesigned, standardized responses to different problems save lives. The basic flowchart leads to a general realization that the processes are not well defined, and that responsibilities are unclear. The hospital management decides to map in more detail some of the processes that are clearly not working well. By doing so, they think BRH will improve its processes, thereby improving their performance in terms of cost, treatment times, and quality.

As already mentioned, it is possible to draw basic flowcharts in different ways. The maps presented in Figure 6.4 and Figure 6.5 are good for following a flow in situations where we are particularly interested in decisions and the way they influence further flow of activities within the process. If decisions and direct links between activities or sub-activities within the process are not that important to you, you could draw a map like the one in Figure 6.6.

This figure also illustrates the process flow by showing the relationship between the activities over time. Just like 6.4, it is a rough picture of the process, representing the same activities. But Figure 6.6 is different from

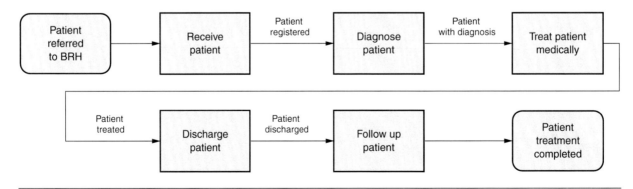

**Figure 6.6**   Holistic patient treatment process—alternate rough version.

Figure 6.4 in that it does not take into account decisions that have to be made at certain stages of the process. If the group is interested only in looking at the content of some or all of these activities, rather than focusing on decisions, this map is more suitable for their needs.

In Figure 6.7, the group starts adding sub-activities included in the main activities. This is done one main activity at a time, and the group starts with the sub-activities belonging to the activity "Receive patient." Figure 6.8 shows what the basic flowchart for this process looks like when this is done for all the main activities.

The group finds it suitable to separate sub-activities involving the patient from activities where the patient does not have to be present. They choose to do this by assigning different colors to the two categories of sub-activities; darker shaded activities can be performed without the patient's presence. The group's reason for doing this is that they find isolating sub-activities involving the patient a good starting point for future improvements in the "patient flow." Sub-activities not involving the patient should support the patient-centered activities, but should not be allowed to obstruct them. The map differentiating between sub-activities with and without the patient is shown in Figure 6.9.

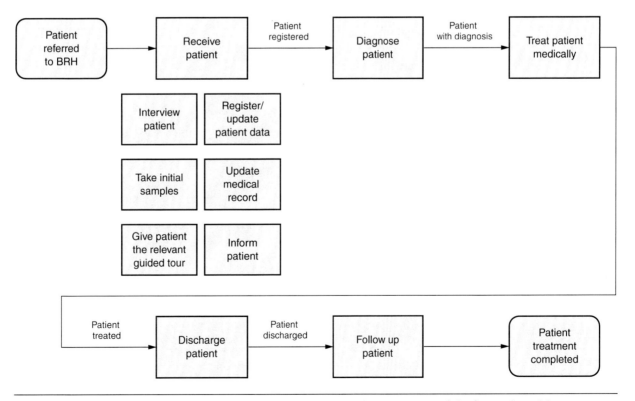

**Figure 6.7**  Holistic patient treatment process—alternate version with sub-activities of the first main activity.

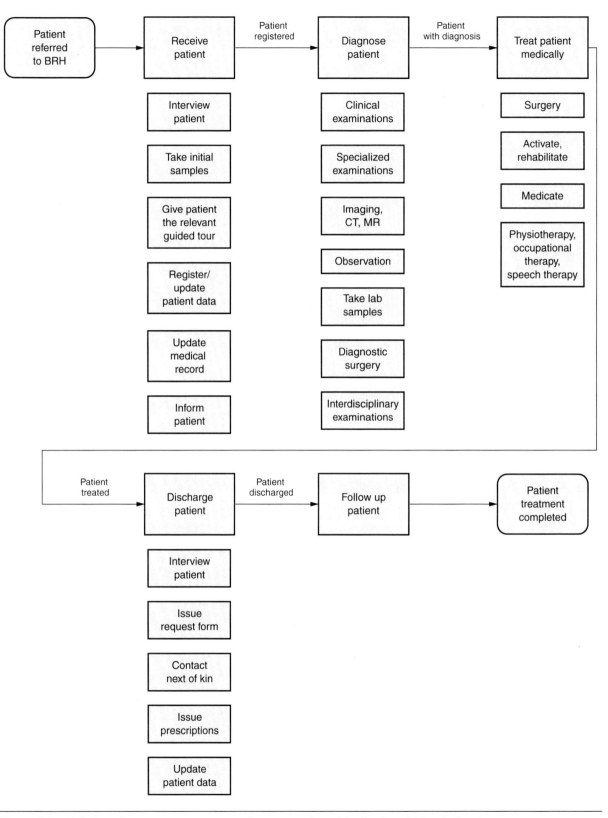

**Figure 6.8**  Holistic patient treatment process—alternate version with all sub-activities depicted.

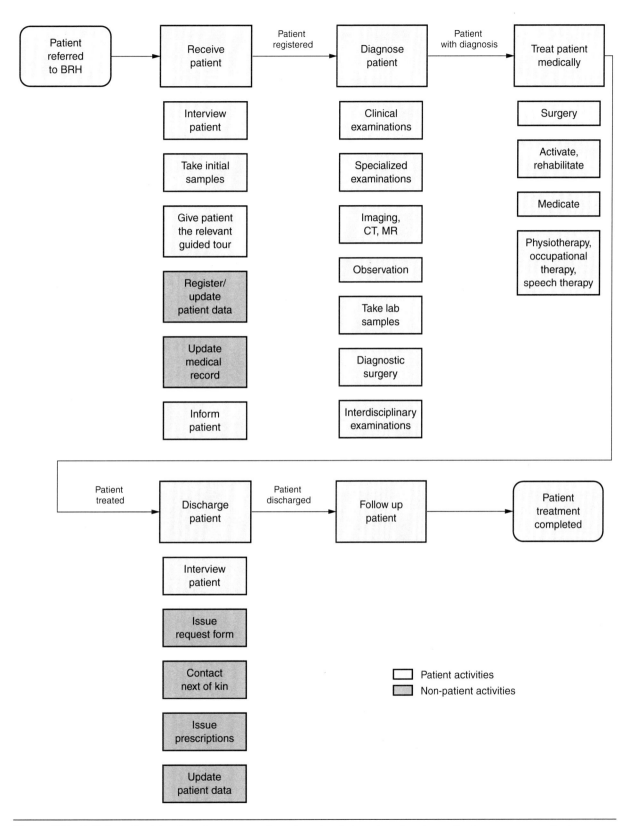

**Figure 6.9**  Holistic patient treatment process—alternate extended version showing patient versus non-patient sub-activities.

## Post-Case Reflection

As you can see from the BRH case, visualizing processes makes it easier to communicate about the different elements of the process. Seeing how activities are related and influence each other helps identify reasons for problems as well as successes. Starting at a high level, you can identify areas that need particular attention and continue by mapping these. The basic flowchart is a good starting point for many process-mapping exercises as it is beneficial to have gone through the basics before you start adding more detailed information. The discussions taking place when constructing a basic flowchart can be as important as the actual map. This is particularly true if the basic flowchart is on a high level, depicting a part of real life that necessarily holds many details.

# 6.4 CHECKLIST

*Checklist for constructing a basic flowchart*

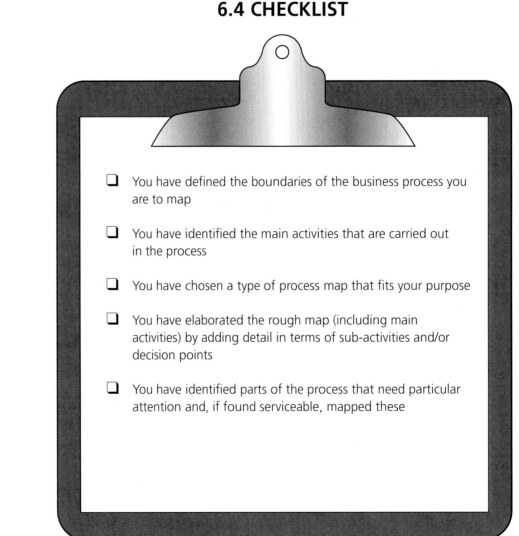

❏ You have defined the boundaries of the business process you are to map

❏ You have identified the main activities that are carried out in the process

❏ You have chosen a type of process map that fits your purpose

❏ You have elaborated the rough map (including main activities) by adding detail in terms of sub-activities and/or decision points

❏ You have identified parts of the process that need particular attention and, if found serviceable, mapped these

# Chapter 7

## Creating a Cross-Functional Flowchart

We have now explained stakeholder maps, high-level process maps, and the basic flowchart. The next logical step is to define the cross-functional flowchart.

### 7.1 WHAT IS A CROSS-FUNCTIONAL FLOWCHART?

Basic flowcharts describe the activities performed in a process. Cross-functional flowcharts additionally tell you who performs what activities and which functional department they belong to. Cross-functional flowcharts are often labeled "swim lane flowcharts," as they can be visualized as swim lanes in a swimming pool (see Figure 7.1). And as we are confident you'll

**Figure 7.1** Swimming pool with swim lanes.

agree, in a swimming pool it is easier to swim in one lane (or a few), than to change lanes often. This is also true for processes.

In our first example of a cross-functional process, a machine operator wants to test a new material for an operation. First he asks his foreman to approve the test (see Figure 7.2).

However, this is a company with a highly hierarchical structure, where one boss after another has to approve a project. Thus the foreman has to ask her boss, the section head (see Figure 7.3).

However, since the company is extremely hierarchical, the section head has to ask his boss, the department head (Figure 7.4).

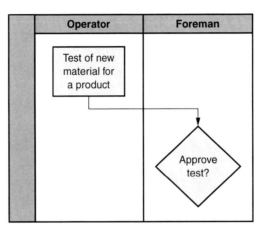

**Figure 7.2**    Machine operator cross-functional flowchart example.

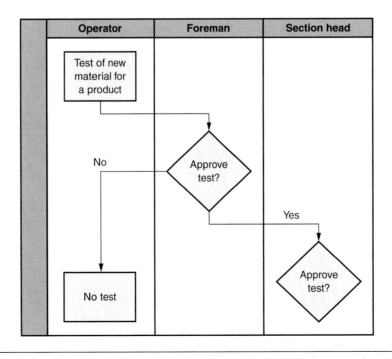

**Figure 7.3**    Machine operator flowchart example expanded to include section head.

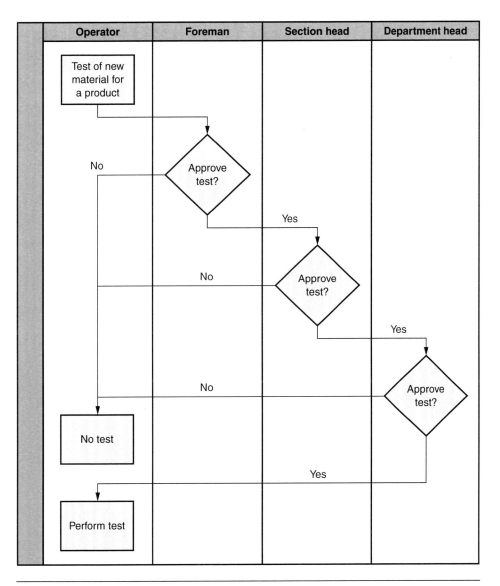

**Figure 7.4** Machine operator flowchart example including department head.

The department head does approve the test, but reduces the amount of material the operator may use. Thus, the test has lost some of its value. The above-mentioned steps illustrate that each activity in the process requires time and resources. Also, information gets lost each time the activity changes lanes. Thus, the operator ends up performing a test that has little or no value. If he could have explained this to the person that can actually approve the test, this could have been avoided. The cross-functional flowchart clearly depicts the steps in the process, and highlights its pros and cons. Another example of a cross-functional flowchart is depicted in Figure 7.5. This example depicts a situation that occurs often these days: a cell phone stops working. As we can see, the cross-functional flowchart clearly depicts the different steps that might occur. Additionally, we see that

*Cross-functional flowchart for a complaint-handling process*

**Complaint handling**

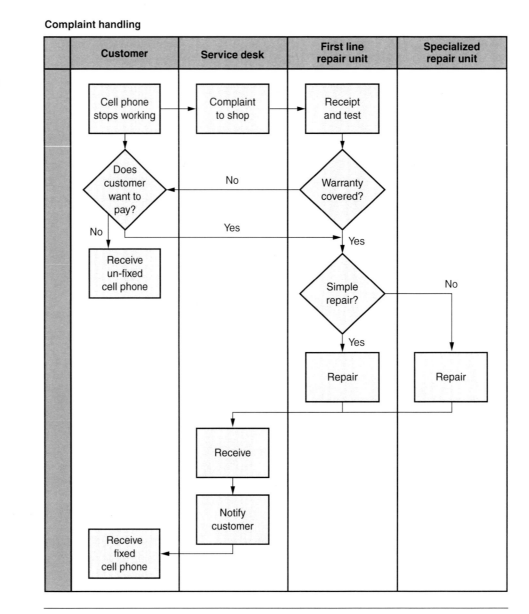

**Figure 7.5**    Example of a cross-functional flowchart for a complaint-handling process.

there are plenty of "lane changes." At each of these interfaces, responsibility shifts and information may be lost. There are several reasons why cross-functional flowcharts are often preferred over basic flowcharts.

*Reasons to prefer cross-functional flowcharts over basic flowcharts*

Cross-functional flowcharts provide a better overview of a process as they depict the different departments involved. This view demonstrates whether the flow is logical or involves a lot of back-and-forth between different units. These insights can provide the basis for improvement opportunities. When a process shifts from one department to another it starts at the back of the line in the new department. This can then cause a delay in the process. The length of the delay will depend on the movement of the queue in the new department. Additionally, as the responsibility shifts from one department to another, vital information may be lost or reduced at the interface.

Back-and-forth interaction between departments that are very different in terms of culture, size, and speed demands that you have a specific focus on these interfaces. In general, this type of flowchart provides a common basis for improving and reengineering the process. It may also be used in benchmarking the process.

It is possible to convert a basic flowchart into the cross-functional type, but it is often more beneficial to start from scratch, even if a basic one already exists.

As described, cross-functional flowcharts have many favorable features. However, among all the types of flowcharts, cross-functional flowcharts are best at depicting a single process flow, whether a physical flow or a flow of information. Flowcharts are not well suited for showing both physical and information flow in the same chart, as the chart easily becomes cluttered. If this is the case, make two variants, one for each type of flow.

*Handling physical and information flow in the same map*

# 7.2 CONSTRUCTING A CROSS-FUNCTIONAL FLOWCHART

The steps in constructing a cross-functional flowchart are based on the same principles as basic flowcharts. We have found the following steps to work quite well:

1. Define the boundaries of the process to be modeled, especially the start and end points. Both start and end points should be well-defined outputs or events.

*The steps in constructing a cross-functional flowchart*

2. Typically starting from the end point, identify the activities that are carried out in the process, along with important outputs, periods of waiting, and so on. For new processes, we strongly advise starting from the end point. However, when reengineering old processes, we often start at the beginning of the process. You should, however, always keep the final outcome of the process in mind, to ensure a customer-oriented process.

3. For each of the activities in the process, determine which organizational unit is in charge of executing it.

4. Construct the flowchart graphically by creating swim lanes or columns that correspond with the organizational units involved, preferably in such a sequence that units in close cooperation are located next to each other.

5. Continue by placing items in sequence and in the proper swim lanes, from the end of the process working backward. In cases of decision points or places where the process branches out, try to orient the diagram so that activities flow in the most logical direction, that is, from left to right and top to bottom.

This procedure will be illustrated for the BRH case shortly.

**Practical advice**

First, some practical advice in designing cross-functional flowcharts:

• When defining departments for swim lanes, don't necessarily take the term *department* literally. In many cases, swim lanes could represent a single person, part of a department, a physical entity, a different company, and so on.

• In cases of disagreement within the mapping team, take the necessary time to resolve any issues so that the final model truly reflects a common understanding. Remember that such disagreements often represent good opportunities for clarifying potential conflicts or misunderstandings.

• When the team is satisfied that the most important elements of the process have been captured and placed in the right sequence, it is a good idea to redraw the process map to increase readability.

## 7.3 CROSS-FUNCTIONAL FLOWCHART FOR BROOK REGIONAL HOSPITAL

According to a study conducted last year by BRH, case cancellations on the day of surgery seem to be a considerable problem. *Case cancellation* means an elective surgical procedure that is not performed on the planned day, despite being on the final operating room schedule. This wastes valuable operating room time. In order to investigate this problem, a team is given the task of mapping the preoperative process. The team chooses to use a cross-functional flowchart, as the transfers between various departments are assumed to have an influence. The team proceeds as follows: The boundaries of the process are defined as starting when a patient receives a letter regarding surgery and ending when that patient is in the preoperative room. For each of the activities in the process, it is determined which organizational unit is in charge of executing it. Finally, the map is drawn without much disagreement, as shown in Figure 7.6.

Feeling comfortable that they shared a common understanding of the process, the team undertook a more thorough study of the process to investigate the high rate of cancellations and figure out the root causes. The real culprit identified was the lack of communication between the laboratory

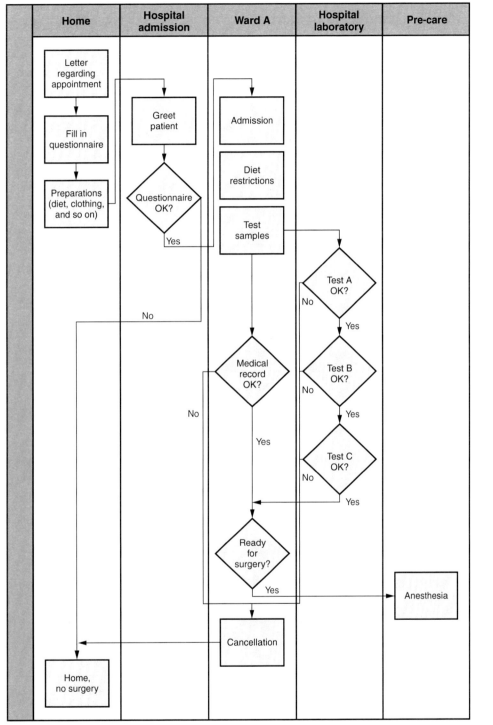

**Figure 7.6** Cross-functional flowchart of preoperative activities at BRH.

and the nursing units. Test results from the laboratory showing a patient not being suitable for surgery did not reach the responsible person at each nursing unit until nearly the last minute. Once this problem was corrected, the cancellation rate went down.

## Post-Case Reflection

Mapping this process demonstrated how valuable it is to reveal responsibilities along the process, and especially how they can shift for different activities. Although this phase of the case focused on the mapping itself, hopefully it also illustrated that mapping can be combined with quality improvement techniques in order to improve the process. The case did not perhaps fully illustrate the potential for improving and reengineering the process based on the cross-functional flowchart, but hopefully did show how a lack of cross-boundary communication can cause difficulties.

## 7.4 CHECKLIST

*Checklist for constructing a cross-functional flowchart*

❑ You have defined the boundaries of the business process you are to map

❑ You have identified the main activities that are carried out in the process

❑ You have identified who is in charge of each of the process activities

❑ You have constructed "swim lanes" corresponding with each department or unit involved in the process

❑ You have completed the map by placing the activities in the right sequence and in the right swim lanes

❑ You have considered whether the process contains illogical shifts of responsibility

# Chapter 8

## Creating a Bottleneck Map/Flowchart with Load Statistics

Mapping processes allows a shared understanding of activities and a basis for more fruitful discussions about performance, where we need to improve, and so on. When we add statistics to the picture, the additional evidence and facts further help us address potential improvements. The level of detail of the statistics can vary from covering all processes and activities to covering only the main processes, or only processes at an aggregate level. At their most advanced, the statistics can form the basis for *activity-based costing* and a completely process-oriented accounting system. However, even without translating the statistics into dollars, they can provide important information about the actual flow through the processes and how the capacity of the different units and people are utilized. This is often referred to as *bottleneck analysis*.

### 8.1 WHAT IS A BOTTLENECK MAP OR FLOWCHART WITH LOAD STATISTICS?

An advantage of flowcharts is the ability to add additional information to your map, such as:

- Time consumed

- The number of items processed

- Costs incurred

- Percentage finished

- Value added

- And so on

*Information that can be added to a process map*

69

There is almost no limit as to what kind of information we could add into the process map, but to keep it readable it is important not to overload it with statistics, especially if the quality of the data is uncertain. The basis for performance measurement of processes normally consists of:

- Flow data or input/output data (number of items)

- Capacity data

- Quality data

*Flow data* represents what is actually happening in the process, and typically covers quantity data for products, parts, components, customers, documents, and so on, that are used in the process.

*Capacity data* refers to the units and people who are performing the different activities; these could include machines, space (for example, warehouses, patient rooms), carriers, equipment, and so on. Such data can also relate to people or special skills and expertise in performing activities. Capacity data is usually presented in terms of available machines, trucks, rooms, people, and so on, often converted into machine hours, man hours, square footage, and so forth, which are easier to translate into costs.

*Quality data* covers information about defects, problems, dissatisfied customers, delays, and so on. From these types of data, we can assess the performance of different activities and the process as a whole. *Performance* means aspects like capacity utilization, costs per activity, how many customers are handled by the salespeople, how many patients are cared for within a unit, and so on.

An example of a process map with statistics is shown in Figure 8.1. The process was mapped as part of a project where a local bank office set out to determine if they had the right capacity of people to serve customers at the general services desk in the front office as compared with the "back office" advisory services and sales.

A basic process map was first drawn showing the different paths of the customers, where some of them were defined in advance through appointments with advisors. "Drop-in" customers were supposed to contact the service desk for information or assessment of needs and eventually be referred to advisors, while others went directly to the cash desk.

The data were collected by counting customers throughout a "normal" week in January. The data were displayed as average numbers of customers per day, which indicated that there was a lack of capacity in the front, causing many customers to be sent to back-office service departments without really needing to see those people. They often had simple requests that could have been solved at the front desk, but the lack of capacity made this impossible. Furthermore, this also consumed valuable time that the

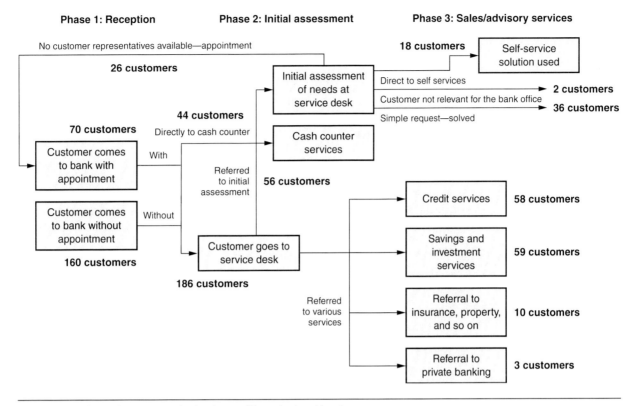

**Phase 1: Reception**     **Phase 2: Initial assessment**     **Phase 3: Sales/advisory services**

No customer representatives available—appointment

**26 customers**

**18 customers**

Self-service solution used

Initial assessment of needs at service desk

Direct to self services    → **2 customers**

Customer not relevant for the bank office    → **36 customers**

Simple request—solved

**44 customers**

**70 customers**    Directly to cash counter

Customer comes to bank with appointment    With

Cash counter services

Referred to initial assessment    **56 customers**

Credit services     **58 customers**

Customer comes to bank without appointment    Without

Savings and investment services     **59 customers**

**160 customers**

Customer goes to service desk

**186 customers**

Referred to various services

Referral to insurance, property, and so on     **10 customers**

Referral to private banking     **3 customers**

**Figure 8.1** Flow of customers in the front office of a local bank.

back-office representatives could have used to follow up with other customers or perform cross-selling activities. The situation was complicated by the fact that the flow of customers varied quite a bit throughout the week and also throughout the day. Deriving a mechanism for flexible allocation of employees between front- and back-office services could help avoid both excesses and shortages of service personnel throughout the office. This exercise was really an example of bottleneck management, where lack of personnel in the front office could cause loss of revenues and failure to provide the required services.

A *bottleneck* is a phenomenon wherein the performance or capacity of an entire process is severely limited by a single component or unit. The limiting point is referred to as the bottleneck. The term is metaphorically derived from the neck of a bottle, where the flow of liquid is limited by its constriction. Bottlenecks are often located on a process's critical path, and limit the throughput of the process. A bottleneck can be a machine, a certain type of employee, their skills, a communication link, a data processing software application, and so on.

Our example from the bank office illustrates how the service desks in the front office could be a bottleneck if the capacity at these desks is not adjusted to the flow of customers. The result of such a bottleneck would

*Bottleneck*

be lines of unsatisfied customers at the service desks. Another negative consequence of the bottleneck was that the processes at the service desks slowed down. Excess capacity in the back office meant that the financial consultants were unproductive, sitting and waiting for the customers that were queuing up in front of the service desks. As described above, this bottleneck was managed by flexing resources between the front and back office, depending on the flow of customers.

In our example, we did not add capacity figures to the different processes; however, the capacity situation was so easy to understand (one person at the service desk plus one more during lunch time) that management could take action without adding these figures. An example from the leisure boat industry shows that there are other situations where the bottleneck is obvious and we don't need many statistics to understand the situation (see Figure 8.2).

*Managing bottlenecks*

The boat manufacturer often was unable to fulfill orders on time, even though their own manufacturing was well scheduled and operated. The problem was narrowed down to the assembly stage, where there were always delays. Not much investigation was needed to identify who was causing the problem: the supplier of windshields, an important module for the boats. This supplier was virtually never delivering on time. Assembly of both the top and important electronic components was dependent on the installation of this module before finishing these other tasks.

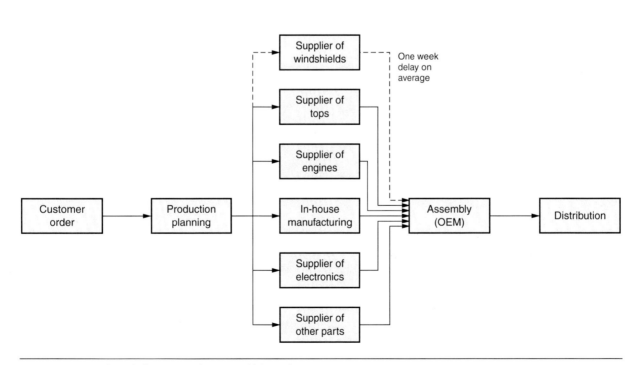

**Figure 8.2**  Bottleneck for a manufacturer of leisure boats.

The problem was that there were no alternative suppliers of windshields, as the current supplier was the only one who could produce them according to the technical specifications. The challenge of managing bottlenecks is even more difficult when they are outside of the company. Apparently the supplier did not have the capacity to satisfy all of its customers. Eventually, the boat manufacturer initiated a joint project with the supplier to improve their production planning and capacity management.

## 8.2 CONSTRUCTING FLOWCHARTS WITH LOAD STATISTICS AND IDENTIFYING BOTTLENECKS

All types of process maps presented in this book can be supplemented with statistics. However, the types of relevant statistics will vary, and the information derived will normally also vary according to the purpose of the process mapping. Load statistics are data that show the strain for different paths through the process, for example, how many products flow through the different manufacturing stations or how many customers choose different treatment paths at a hair dresser. Typically, charging flowcharts with statistics is done as follows:

1. Determine the purpose of adding load statistics to the process map.

2. Clarify what kinds of information you need to better understand the process and its performance.

3. Identify the data that could give you this information.

4. Identify the sources of these data.

5. Assure the quality of the data.

6. Present the statistics in a way that enables analysis.

*Steps in adding statistics*

There is always a purpose for drawing a process map, and the first step in adding statistics to the map is to elucidate to what extent this will improve the explanatory impact of the process map. There are almost no limitations as to what kind of statistics can be tagged to the process map, but to keep the process map easy to understand we should be careful not to add statistics that increase its complexity without adding value to the process map.

There are typically large amounts of data available in most organizations, for example, accounting data, data or information processed for governmental bodies, or data automatically collected and stored in different machines. Before creating new data sources, you should carefully review

*Huge amounts of data exist in all organizations*

existing sources to see if they could be used and whether the data are of the required quality.

*New technologies*

New technologies allow an increasing amount of data to be accessed through electronic devices. For example, in most cars the onboard computer has changed the work of mechanics dramatically as they now have access to a wide range of diagnostic data without turning a single bolt. RFID (radio frequency identification) technology is an example of an emerging technology enabling measurement of processes using sensors and readers. Work processes can also be described using such data. Another example is the counting of "hits" or operations in a Web-based logistics system.

*Manually collected data might be necessary*

In our example from the local bank office, data from the queuing system could have provided important statistical material. However, as this machine could not give information about what happened to the customer after leaving the service desk, a manual counting was performed.

If the existing data sources are not good enough, new sources might be required. However, it could be both expensive and time-consuming to establish new sources that yield high-quality data, and the benefits of these statistics must justify the effort.

*Presenting the statistics*

There is no uniform way of presenting the statistics, but to employ the visual strength of process maps, we strongly recommend presenting at least the basic statistics in the maps. There are three ways of presenting the statistics: a) we could tag both the arrows (input/output) and the process boxes with (capacity) data, b) we could have all the relevant statistics in the process boxes, or, of course c) a mix of these two ways of presenting statistics. Some examples are shown in Figure 8.3.

*Presenting statistics in tables with references to the flowcharts*

In some cases, adding statistics to the process map could be difficult, and important information could "drown" in a complicated picture. An alternative method of showing the statistics is to present them in a table, but with clear references to the flowchart, as illustrated in Figure 8.4.

Most statistics and information relevant for analyzing and managing bottlenecks can be added to such tables, for example, a detailed presentation of critical resources, how and when they turn into bottlenecks, and how they could be managed.

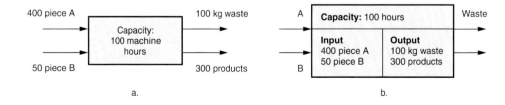

**Figure 8.3**   Presenting statistics in the process map.

| Process | | | | | | |
|---|---|---|---|---|---|---|
| Activity | | Flow data | | Capacity | | Bottleneck management: |
| No. | Name/ description | Input | Output | Critical resource | Description | How could capacity and flow be managed |
|  |  |  |  |  |  |  |
|  |  |  |  |  |  |  |
|  |  |  |  |  |  |  |

**Figure 8.4**   Template for table displaying statistics in association with a process map.

Bottlenecks may be identified as a result of statistics being added to the process map, but may also be identified as the result of the data providing indications of the presence of bottlenecks but where identification requires a more detailed analysis. This situation would imply the following steps:

1. Identify accumulation of items, queues, delays, and other indicators of bottlenecks along the process.

2. Determine where in the process the problems appear most frequently.

3. Analyze the causes for the problems, for example, capacity problems or quality issues.

4. Collect statistics if necessary.

5. Identify and carry out appropriate actions to manage the bottlenecks.

*Steps in identifying bottlenecks without statistics*

Identifying bottlenecks often means applying the statistics to determine whether there are any areas with an obvious lack of required capacity to handle the input and produce the output that is needed for the next activity or process. In the example in Figure 8.3, we also need to know how many products the machine can produce per hour to determine whether this machine might be a bottleneck. In this example, the machine is able to produce three products per hour, thus the 300 products per day required. However, if the machine were able to produce only two products per hour, it would probably represent a bottleneck as it would not produce the 300 products required in the 100 machine hours available. Another result of this bottleneck would be a queue of the input pieces A and B, an unwanted stock of unfinished goods. Our example from the local bank office also illustrates how adding statistics to the flowcharts gives us indications of possible bottlenecks. One obvious benefit of adding statistics to the flowchart is that it gives us early warning of bottlenecks that could appear if the flow increases or capacity is reduced. This would enable us to act to prevent problems and dissatisfied customers. The steps in using statistics to identify bottlenecks are:

*Steps in identifying bottlenecks using statistics*

1. Add capacity and flow data to the process map.

2. Identify variation in flow (and capacities) over time.

3. Determine if the capacity in each step is sufficient to handle the flow at the different stages of the process in a normal situation.

4. Also consider whether the capacity is sufficient to handle load variations.

5. Define any bottleneck where the capacity is not sufficient to handle the flow.

6. Identify and carry out appropriate actions to manage the bottleneck.

Very often, you will find yourself in a situation where it will be too demanding on resources to build capacities to handle peaks in the flow. In these cases, you need to define acceptable levels of queues and the capacity required.

*How to manage bottlenecks*

The next step is to find ways to manage the bottleneck, either by removing it or reducing the negative consequences. In our machine example (Figure 8.3), we could try to improve the machine's capacity from two to three products per hour, or we could increase the number of machine hours, for example, by adding more machines or an extra shift to the process. Or perhaps we could change the process completely so that we would not have to rely so much on this machine.

A third option could be to reduce the required number of products from 300 to 200 per day. This would be a very good option if the production required varies during the year, and the 300 figure represents a peak in the demand and not an average. If we could smooth out the demand for products, the machine might not represent a bottleneck any more. This is often termed *balanced flow* or *balanced processes*.

*Practical advice*

Next we will illustrate how bottleneck identification through statistics added to process maps could be useful for BRH, but first some practical advice and hints about adding statistics and identifying bottlenecks:

• Don't overload the process map with statistics. You could easily lose the overview it provides and obscure other important aspects. You can always make several process maps with different statistics added to each instead.

- The process map could be useful as a framework or tool for registration of data, almost like a check sheet.

- Unreliable or poor-quality data are often considered as fact, or as more valuable than they actually are. Make an effort to ensure data quality.

- Try to identify data that could be used as indicators for several aspects of the process.

- Bottlenecks don't necessarily have to be removed. Removing a bottleneck may create others that are even harder to manage. The most important factor is to be aware of the bottleneck and adjust activities so that the bottleneck doesn't cause too much trouble.

- Bottlenecks can often be identified without adding statistics. A visual assessment and common sense will often be enough.

## 8.3 UNDERSTANDING THE BOTTLENECKS OF BROOK REGIONAL HOSPITAL

After an impressive effort by the whole organization, BRH has now entered a phase where they might well be described as a process-oriented hospital organized along the patient pathways, and where "functional silos" are more or less history. Also, the top management of PCH has expressed their acknowledgment of these efforts. Now is the time to benefit from this effort, harvesting performance improvements in terms of reduced throughput time, reduced costs, and better quality of services.

But the results are somehow disappointing, especially for some of the processes and patient groups where the benefits of the process orientation seemed most obvious, such as orthopedic services. As a result of the new strategy, aiming for a higher degree of specialization and a focus on the most profitable patients, the orthopedic services were highlighted. Orthopedic services were considered to have a large potential for streamlining, as the patient groups tend to be more homogenous, often with less complex diagnoses and treatments.

*Disappointing performance*

The first action to be taken was to separate the acute and elective processes, since the "interference" of acute patients made it almost impossible to make plans for the elective operations and care. This interference resulted in battles for rooms, operating theaters, surgeons, and so on. The elective patients were almost always downgraded and often had to return home from

*Separate patient pathways for acute and elective patients*

the hospital, to be rescheduled at a later time. These patients were dissatisfied and expressed it loudly, not only to BRH but also to the local media.

The key to the new strategy was to split the orthopedic activities into two separate processes: acute and elective. Each process was given dedicated resources such as personnel, rooms, and equipment. The sustained high performance of the acute process demonstrated a capacity for delivering high quality services. The elective process still suffered from less-than-desired performance. There were still delays, with patients and tasks accumulating, particularly in the initial stages. The paradox was that for other stages of the process, there seemed to be excess capacity, even though the orthopedic surgery superintendent had warned about a lack of capacity, especially for operations. The management thought that this situation might indicate one or more bottlenecks in the process, and a task force was established to investigate the elective orthopedic process. Through analysis of the process and discussions with the people involved, the team got the impression that the problems were related to imaging and tests, and also in post-operational activities such as recovery. This could be seen in the basic process map (see Figure 8.5). However, definite conclusions were difficult to draw and more statistical evidence was collected.

A great deal of data is produced within a hospital: statistical reports to authorities, claims reports to insurance companies, accounting data, and so on. Registrations and records from patient journals were also considered

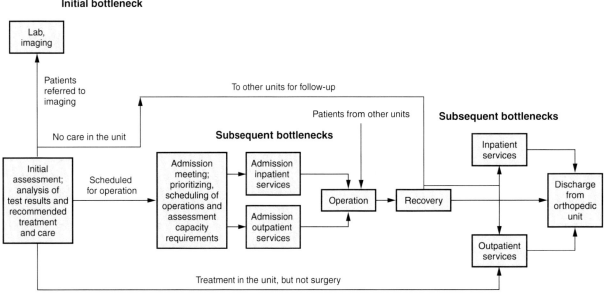

**Figure 8.5**  Map of elective orthopedic process.

relevant data sources. To assure the quality of the data, the different registers were compared, which revealed almost no consistency. The team decided to carry out its own collection of data for the analysis during one month. Both capacity and flow data were collected, extrapolated into one-year statistics, and added to the process map, as illustrated in Figure 8.6.

One bottleneck was obvious: the laboratory did not have the capacity to perform the tests required in a timely manner, resulting not only in queues at the lab, but also causing a problematic situation at the initial assessment, where people without the relevant tests and x-ray results often had to wait or be rescheduled. Considering the resources and capacity present in the lab (which is a shared service at the hospital), they actually did quite a good job. But increasing the capacity was not an option, and the task force had to look for ways of reducing the test referrals. Interviews were carried out, and they discovered that a lot of unnecessary tests were performed as a result of poor planning and sloppy paperwork among the referring doctors. Tests were requested where diagnoses were already 100 percent certain, and x-ray results often "disappeared" and the procedures had to be performed over again. These were not only problems within the orthopedic processes, but existed hospitalwide.

Improved routines and pressure from management changed the situation completely during the next two months, and test referrals were reduced by 20 percent. The bottleneck at the lab was removed, but after only a couple of months, the task force was again called in for a meeting. It was obvious

*Removing one bottleneck, creating another*

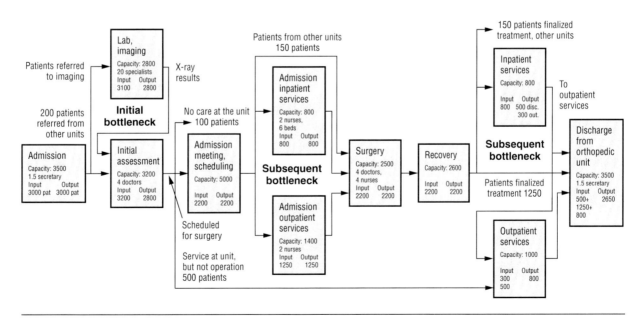

**Figure 8.6** Map of elective orthopedic process with annualized statistics.

that a new bottleneck had surfaced after the removal of the lab bottleneck had increased the patient flow by 10 percent. This increase created capacity constraints at the operating theaters and in certain rooms and beds for inpatient services. Again, the management group was not very keen on investing in more capacity to remove the bottleneck. Their decision was to increase the capacity for outpatient services, that is, for those who could go home after surgery, enabling more patients to take this much less resource-demanding route.

The BRH management group was very satisfied with the new insight gained from the statistics and the bottleneck analysis. They understood that the statistics they had were not very useful for performance measurement and for managing their processes, and initiated a program for improved process measurement. The objectives were defined as: improving the quality of existing statistics, avoiding registration redundancy, and enabling process-oriented statistics.

*Improvement without major capacity increases*

The hospital is now quite proud of the orthopedic processes and their performance, and is curious about how they compare with other hospitals. The new statistics improved the basis for performance benchmarking with other hospitals in the PCH group, and BRH also joined a common interest group (benchmarking club). This showed that their performance was now in the top five percent within the group.

The new statistics provided evidence to support the PCH management's impression: BRH, its employees and management group, had made a significant effort to continuously improve, and had succeeded. BRH was a hospital worthy of future investments, inducing PCH to agree to the new pediatrics wing!

## Post-Case Reflection

In this phase of the case, we have illustrated the importance of adding statistics to process maps to better analyze performance, bottlenecks, and so on. This knowledge allows assessing current performance and documenting the results of improvement efforts.

There will typically be large amounts of data within your organization, and they might be relevant as sources for process statistics. However, there will always be a trade-off between the cost of processing high-quality data and the benefits reaped from the statistics and information they provide. As we have seen from the examples in this chapter, knowledge of possible bottlenecks might be gained without the use of statistics, but in the long run a process-oriented organization should implement continuous process measurements.

# 8.4 CHECKLIST

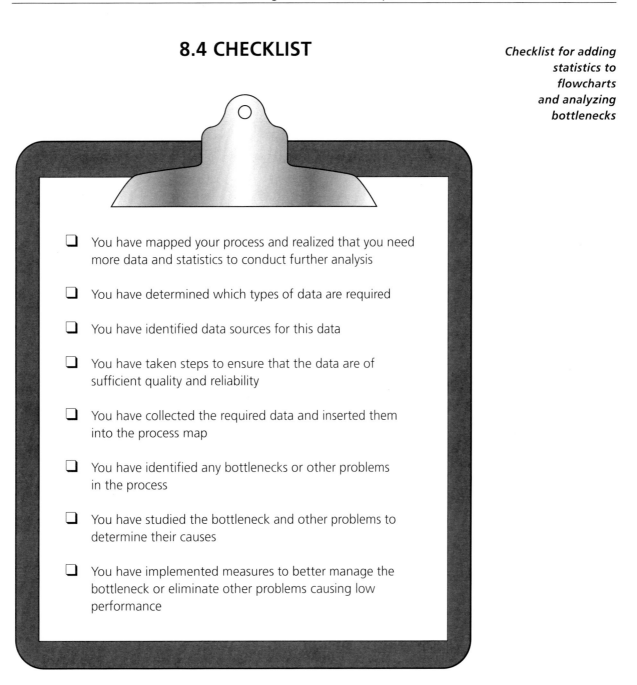

❑ You have mapped your process and realized that you need more data and statistics to conduct further analysis

❑ You have determined which types of data are required

❑ You have identified data sources for this data

❑ You have taken steps to ensure that the data are of sufficient quality and reliability

❑ You have collected the required data and inserted them into the process map

❑ You have identified any bottlenecks or other problems in the process

❑ You have studied the bottleneck and other problems to determine their causes

❑ You have implemented measures to better manage the bottleneck or eliminate other problems causing low performance

# Chapter 9

---

# Using the Process Maps

Throughout the previous chapters, we hope we have achieved what we set out to do with this book: present easy-to-understand instructions and guidelines on how to map work processes. Only you can judge whether we have been successful, but being bold enough to assume that we have, we could very well close the book without this chapter. We will instead reiterate some advice touched on throughout the book and also emphasize the need to *use* the process maps, not just file them away for eternity.

## 9.1 SOME FINAL ADVICE

Unlike the first edition of this book, we have discussed several different types of process maps, and we have also shown examples of maps with quite different appearances. This has been quite deliberate on our part as this reflects more closely how things are actually done in different organizations. If you compare process maps from two similar companies, you will often find vastly dissimilar styles, symbols, levels of detail, and orientations. The key point is that your organization should find a way of drawing process maps that you feel comfortable with, that is well known throughout the organization, and that is able to convey the messages you seek to include in the maps. We therefore hope you will view the process maps shown in this book merely as examples of how they can be constructed.

*Find your own way of mapping processes and stick to it*

We also owe you an apology of sorts. When writing this book, we have used examples that are suitable for presentation to our readers, meaning they contain logical sequences of activities, have easily identifiable start and end points, are suitably complex to fit on the page, and so on. From painful

*Real processes are more complex than our maps indicate*

experience, we know that you will very likely run into situations where you're pulling your hair out and screaming because your map does not look anything like what we showed in the book! Real-life work processes are often inherently complex, without a straightforward main sequence of activities. There may be several parallel paths a process can take, or a process may change over time, either gradually or radically from day to day, depending on who staffs it and which customers are being served. You might even find that parts of the organization can at best be said to constitute a collection of people that by coincidence manage to serve their customers, without any discernable processes to show for it. This is completely normal!

Typically, the more often a process is repeated and the longer it has been in place, the more it tends to be standardized and follow a well-defined sequence of activities. Processes that occur less often, that require improvisation, or that involve a strong component of "knowledge work" are usually inherently more difficult to document. And this is okay, you should not try to force such areas of the organization into a mold they will not fit into anyway. Mapping what is going on is still useful; it may tell you that everything must be improvised every time, it may reveal interfaces between people and tasks so far not well understood, or it may demonstrate that parts of the process can be standardized whereas others must be allowed to run more freely. This knowledge is also useful.

*Don't let process maps freeze your work processes*

A word of caution is also in order: there is a tendency for process maps to contribute to making process designs more static than they ideally should be. The more effort that is put into creating a detailed process map, perhaps publishing it on intranet pages, or even building your management control system around the mapped process, the more tempting it is to let the process remain unchanged for some time. Any improvement in the process means that the process map must be redrawn and these changes updated wherever the map appears. Two lessons can be learned from this. First of all, keep process maps in a format that allows easy updating; there is perhaps no worse excuse for not implementing obvious process improvements than wanting to avoid cumbersome process map revisions. Secondly, be aware of this phenomenon and make sure that process maps are used as a tool for creating better process insight. The moment you let your process maps represent the "official" process design truth, you've allowed the process maps to dominate your work processes. A map is a representation of the real world, not the other way around, and if there are discrepancies between the two, the error is always in the map! Work processes will inevitably change over time, sometimes slowly and incrementally, sometimes exponentially. This is simply the way of things. Allow them this freedom, and update the process maps accordingly.

You should also keep in mind that despite how powerful process maps can be, there are many aspects of a work process that cannot be captured and depicted in a process map. The organizational culture determines to a great extent how well the process is performed; the same goes for the competence levels and skills of those conducting its different steps. This means that you will also need to consider such issues when discussing process performance and possible ways to improve it, in addition to the flow of activities and responsibilities.

Finally, acknowledge that mapping work processes is a skill. You can't expect people who have never been involved in such an exercise to perform well from the very first go. Train people in work process mapping, and allow them to gain experience and sharpen the skill over time. Perhaps most importantly, ensure that a certain portion of the organization has a working knowledge of process mapping. Once you start applying this technique, chances are that you will regularly see the need to undertake some mapping. Having an inherent capacity for this within the organization is extremely useful.

*Hone the skill of mapping work processes*

## 9.2 PROCESS MAP USAGE

It would be negligent on our part to leave you with the impression that process maps are objectives to pursue for their own sake. Our position is clear: while we acknowledge their importance, both the exercise of constructing the process map and, not in the least, what you do with them afterwards, are much more important. Process maps are primarily a means to an end, or several ends in fact. As we discussed in Chapter 1, process mapping can serve many different purposes, but you will not realize these purposes just by mapping processes. In many cases, the process map represents merely a starting point for further analysis; often the map constitutes the whole or part of the decision support basis for resolutions to implement process improvements, acquire new equipment, hire people, reorganize, and so on. The full benefits of process mapping will only be achieved when such decisions have been made and implemented.

*Put the process maps to use!*

On the other hand, we have been quite determined that this book would not be stretched into detailing the ensuing stages of process improvement. For the book to remain focused and easy to use, we decided to restrict it to the core activities of process mapping. Besides, there are numerous titles available that deal extensively with different types of process analysis, business process improvement, and change management, to mention a few relevant topics. ASQ Quality Press carries many recommended titles, so please browse their catalog, search the Web, or consult your library for further reading.

In Figure 9.1, we have tried to illustrate the phases involved in working with processes. The diagram assumes an organization with no initial

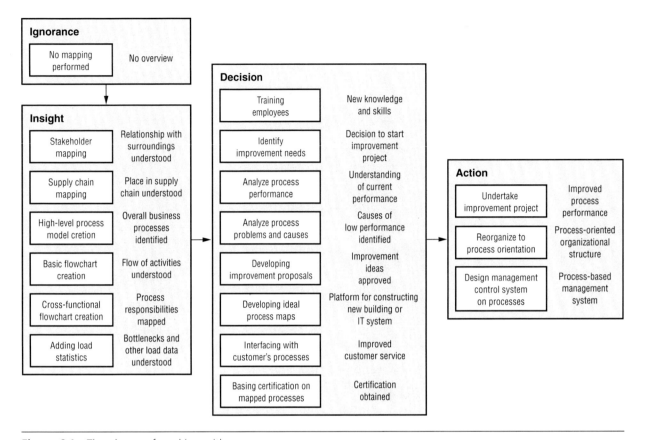

**Figure 9.1**   The phases of working with processes.

conscious understanding of work processes. Having undertaken several or many different types of process mapping, the organization will reach a stage of insight: Where do we fit into the bigger scheme of things? Which work processes does the organization run? How are these performed? and so on. Stopping here, however, is claiming victory too soon. The next phase is to put the process maps to use, for one or more of the different purposes outlined. This typically leads to decisions to implement changes and improvements. Finally, when these have been implemented, resulting in improved performance, a truly process-oriented organization, or possibly implementing a process-based management system, the mapping efforts have really paid off.

In presenting this diagram, we are certainly not saying you should use your process maps for all of these possible applications. But we do recommend that you make a conscientious effort to reap more benefits from them than have been achieved during the actual mapping exercises. Discuss how they will be used, make plans for future projects where they can be utilized, and follow up on the mapping exercise; process maps become obsolete rather quickly.

# Appendix

# Process Definition Template

| Define the *end point* of the process | ✏️ Notes |
|---|---|
| **Advice** | |
| The process end point is a product or service that is created by the process. This is the final output of the process.<br><br>End points should be expressed as a noun and a verb for example, "treatment completed," "cell phone fixed," "self service solution used." By expressing the end point as a noun and a verb you are forced to consider the product and the action related its condition. | |

| Define the *starting point* for the process | ✏️ Notes |
|---|---|
| **Advice** | |
| The starting point for the process. This is the initial input to the process.<br><br>Starting points should be expressed in the same format as end points, as a verb and a noun. For example "patient referred to hospital," "cell phone not working," "customer arrives in bank." | |

## Define the *parallel processes*

### Advice

**Notes**

You have already started to define the process boundaries by stating the end point and starting point for your process. To further define the process it can be helpful to compare processes that run in parallel to it.

Ask yourself: What are the processes running in parallel to the process I/we are mapping? Are there any activities from parallel processes that should be included because they are shared or connected in any way to the process I/we are mapping?

You may find it necessary to change the boundaries to parallel processes later.

## List the *customer(s)*

### Advice

**Notes**

The customer(s) are person(s) who use what is available from the end point of your process. The output from your process is the input to the process of your customer(s).

Use names (individual or organizational), roles, job titles, or combinations of these when listing your customers.

Remember that customers can be found within your organization as well as externally.

## State *customer requirements*

### Advice

**Notes**

What do the customer(s) require from the output of your process? Express requirements in terms of, for instance, timeliness, quantity, user-friendliness, appearance, and cost.

Often requirements are related to the interface between your process and a following process.

Make sure to distinguish between customer requirements that are absolute and the ones that are not.

## List the *process participants*

### Advice

**Notes**

The process participants are the people actually performing the steps of the process.

Pay particular attention to managerial roles, support functions, and subcontractors. Ask yourself: is this person actively involved in performing any specific element of the process?

Use names, roles, job titles, or combinations of these when describing process participants.

## Identify the *process owner*

### Advice

The process owner is the individual who is responsible for the process. Often this will be the head of the department in the organizational unit where most of the process is performed.

The process owner is not only a key decision maker, but also a good source of information about the process.

Notes

## List the process *stakeholders*

### Advice

A stakeholder is a party who affects, or can be affected by, the process itself or its outcomes.

Stakeholders can be individuals, groups, companies, or government agencies. Examples of stakeholders are shareholders, employees, media, patients, competitors, and the external environment.

Remember to look for stakeholders both within and outside your process boundaries.

Use names, roles, job titles, or combinations of these when listing the process stakeholders.

Notes

## List the process *suppliers and their contributions*

### Advice

A supplier is anyone contributing with goods or services that are necessary in order to carry out the process.

Names of organizations, departments, and functions are common ways of describing suppliers.

List the suppliers together with their contributions to the process. Examples: "Test results from the hospital laboratory," "Financial support from the research council," "Building materials from the lumber mill," "Sales reports from the sales department."

Notes

# Recommended Reading

As you might have noticed while reading this book, there are no references in the text, unlike more academically oriented volumes. This has been a deliberate decision on our part, as we wanted to keep the book as practitioner-oriented as possible. However, we do realize that many readers will want to pursue different topics further, and we'd be negligent not to give some advice on where to look for additional material. Thus, this section contains a list of recommended reading that we have enjoyed over the years (or in some cases written and hope others have enjoyed).

Altshuller, Genrich S. *40 Principles: TRIZ Keys to Technical Innovation,* translated by Lev Shulyak and Steven Rodman. Worchester, MA: Technical Innovation Center, 1997.

Ammerman, Max. *The Root Cause Analysis Handbook: A Simplified Approach to Identifying, Correcting, and Reporting Workplace Errors.* New York: Quality Resources, 1998.

Andersen, Bjørn. *Business Process Improvement Toolbox,* 2nd ed. Milwaukee: ASQ Quality Press, 2007.

Andersen, Bjørn, and Tom Fagerhaug. *Root Cause Analysis: Simplified Tools and Techniques,* 2nd ed. Milwaukee: ASQ Quality Press, 2006.

Bertels, Thomas, editor. *Rath & Strong's Six Sigma Leadership Handbook.* Hoboken, New Jersey: John Wiley & Sons, 2003.

Burr, John T. "The Tools of Quality: Part I: Going with the Flow(chart)." *Quality Progress,* June 1990.

Davenport, Thomas H. *Process Innovation: Re-engineering Work Through Information Technology.* Boston: Harvard Business School Press, 1993.

De Feo, Joseph A., and William W. Barnard. *Juran Institute's Six Sigma Breakthrough and Beyond.* New York: McGraw-Hill, 2004.

Fellers, Gary. *The Deming Vision SPC/TQM for Administrators.* Milwaukee: ASQC Quality Press, 1992.

Galloway, Dianne. *Mapping Work Processes.* Milwaukee: ASQC Quality Press, 1994.

Gitlow, Howard, Alan Oppenheim, and Rosa Oppenheim. *Quality Management: Tools and Methods for Improvement.* Burr Ridge, Illinois: Irwin, 1995.

Hammer, Michael, and James Champy. *Re-engineering the Corporation: A Manifesto for Business Revolution.* New York: Harper Business, 1993.

Harrington, H. James. *Business Process Improvement: The Breakthrough Strategy for Total Quality, Productivity, and Competitiveness.* New York: McGraw-Hill, 1991.

Hunt, Daniel. *Process Mapping: How to Reengineer Your Business Processes.* New York: John Wiley & Sons, 1996.

Imai, Masaaki. *Kaizen: The Key to Japan's Competitive Success.* New York: Random House Business Division, 1986.

Kubeck, Lynn C. *Techniques for Business Process Redesign: Tying It All Together.* New York: John Wiley & Sons, 1995.

Kume, Hitoshi. *Statistical Methods for Quality Improvement.* Tokyo: Association for Overseas Technical Scholarships (AOTS), 1992.

Mizuno, Shigeru, editor. *Management for Quality Improvement: The 7 New QC Tools.* Cambridge, MA: Productivity Press, 1988.

Peppard, Joe, and Philip Rowland. *The Essence of Business Process Re-engineering.* Hemel Hempstead, England: Prentice Hall, 1995.

Robson, George D. *Continuous Process Improvement: Simplifying Work Flow Systems.* New York: The Free Press, 1991.

Scholtes, Peter R. *The Team Handbook: How to Use Teams to Improve Quality.* Madison, WI: Joiner, 1988.

Straker, David. *A Toolbook for Quality Improvement and Problem Solving.* London: Prentice-Hall, 1995.

Swanson, Roger C. *The Quality Improvement Handbook: Team Guide to Tools and Techniques.* London: Kogan Page, 1995.

Tague, Nancy R. *The Quality Toolbox,* 2nd ed. Milwaukee: ASQ Quality Press, 2005.

Wilson, Paul F. *Root Cause Analysis Workbook.* Milwaukee: ASQC Quality Press, 1992.

# Glossary

**alternative path**—A path through a flowchart comprising one or more optional tasks off the mandatory primary path; preceded by a decision diamond.

**basic flowchart**—A simple map depicting the activities, with inputs and outputs, performed in an individual process.

**bottleneck**—A situation where the performance capacity of an entire process is severely limited by a single component/unit.

**capacity**—The amount of work a "production unit," whether an individual or group, can accomplish in a given amount of time.

**capacity data**—Data referring to the people and units who are performing the different activities. These could be machines, space (for example, warehouse, patient rooms), carriers, and so on. Capacity data also refers to special skills and expertise.

**consensus**—Agreement, harmony, compromise. A group decision that all members agree to support, even though it may not totally reflect their individual preferences. Consensus is possible when diverse points of view have been heard and examined thoroughly and openly.

**core process**—Core processes are the fundamental activities or groups of activities that are so critical to an organization's success that failure to perform them will result in deterioration of the organization. These are typically processes that directly touch the organization's customers, reflect the major cost drivers in the organization, or are on the critical path in the service chain. Core processes are most often found within the customer/consumer lifecycle in an organization, from the first interaction of a consumer with an organization to the last interaction in the relationship.

**cross-functional flowchart**—A process map that uses "swim lanes" to supplement the basic flowchart with information about who (persons, departments, and so on) is responsible for each activity, thus allowing the identification of illogical assignments of responsibility.

**customer**—The person or persons who use your output—the next in line to receive it. Whether your customers are internal or external to your organization, they use your output as an input to their work process(es).

**decision diamond**—A diamond shape in a flowchart that poses a question and signals either an alternative path or an inspection point.

**elective patients**—Planned patient treatment.

**facilitator**—The role of the leader of a group process.

**flow data**—Flow data represent what is actually going into the activities, typically products, parts and components, documents, and so on, that are objects used in activities.

**input**—The materials, equipment, information, people, money, or environmental conditions that are needed to carry out the process.

**inspection point**—A pass/fail decision, based on objective standards, to test an output in a process. Signaled by a decision diamond with two or more paths leading from it. May lead to a rework loop (step) or to a do-over loop.

**internal value chain**—The internal value chain consists of a set of primary work processes that deliver various outputs to the next process in line, as well as processes that support the primary, value-adding processes. (See *value chain*.)

**macro process**—Broad, far-ranging process that often crosses functional boundaries (for example, the communications process or the accounting process). Just a few or many members of the organization may be required to accomplish the process.

**mandate**—An authorization, command, or instruction to perform certain activities. For example, an authorization, including the scope, for improving certain processes.

**mapping**—The activity of creating a flowchart of a work process showing its inputs, tasks, and activities, in sequence.

**micro process**—A narrow process made up of detailed steps and activities. Could be accomplished by a single person.

**OEM**—Original equipment manufacturer; the company that produces the products sold to the value chain's end customers.

**output**—The tangible product or intangible service that is created by the process; that which is handed off to the customer.

**parallel process**—A process executed by someone (or something) else that occurs simultaneously (concurrently) with the primary process. May or may not be part of the primary process.

**parking lot**—Area for storing ideas that are not suitable for actual discussion during group work, but are relevant for other purposes. The parking lot is a documentation of these ideas.

**patient pathway**—The linkages of treatment and service activities related to patients.

**primary process**—A sequence of steps, tasks, or activities that converts inputs from suppliers into an output. A work process adds value to the inputs by changing them or using them to produce something new. (See *core process*.)

**process**—A collection of activities that takes one or more kinds of input and creates an output that is of value to the customer.

**process boundaries**—The first and last steps of the process. Ask yourself, "What's the first thing I/we do to start this process? What's the last step?" The last step may be delivery of the output to the customer.

**process map**—A graphic representation of a process, showing the sequence of tasks; uses a modified version of standard flowcharting symbols.

**process owner**—The person who is responsible for the process and its output. The owner is the key decision maker and can allot organization resources to the process participants. He or she speaks for the process in the organization. That is, if someone says, "How come those California people aren't selling enough equipment?" the process owner—probably a district sales manager on the West Coast—would have to come forward to answer.

**process participants**—The people who actually perform the steps of the process—as opposed to someone who is responsible for the process, such as the process owner/manager. For example, if you use subcontractors to produce the product, and you don't actually do the work yourself, the subcontractor is the process participant.

**quality**—Conformance to customer needs, wants, and expectations—whether expressed or unexpressed. Fitness for use.

**requirements**—What your customer needs, wants, and expects of your output. Customers generally express requirements based on the characteristics of timeliness, quantity, fitness for use, ease of use, and perceptions of value.

**rework loop**—The result of a failed inspection point. A rework loop adds steps to the process and generally leads back to the inspection diamond.

**secondary process**—Processes that are needed to enable core processes to run, but are not directly involved in satisfying customers.

**secretary/recorder**—The role of the person responsible for documenting the discussion and the output from a group process.

**stakeholder**—A process stakeholder is someone who could be a supplier, customer, or process owner, but also could be someone who has an interest in the process and stands to gain or lose based on the results of the process. Most processes have a number of stakeholders, such as senior managers from other departments or even government agencies.

**stakeholder map**—A graphical presentation of how the organization is located within a larger set of surroundings and which links it has to different actors, internal and external.

**subprocess**—The smaller steps that comprise one process step; the next level of detail. Has all the same characteristics of a primary process, such as decision diamonds, parallel processes, or inspection points.

**supplier**—The people (functions or organizations) who supply the process with its necessary inputs.

**supply chain**—See *value chain*.

**support processes**—See *secondary process*.

**value chain**—The chain of organizations, linked in supplier–customer relationships, that join forces to deliver a service or product to the end customer.

**value chain map**—A process map that defines the organization's position in the larger chain of supplier–customer relationships that form the chain from raw material suppliers to end vendors of products or services to the final customer, and shows which individual work processes are performed in the organization and how these are linked together in chains or networks inside the organization.

**value network**—Complex sets of social and technical resources that work together via relationships to create economic value. Value networks are often represented by several value chains. (See *value chain*.)

**value-added step**—A step that contributes to customer satisfaction. A customer would notice if it were eliminated.

**value-creating process**—See *core process*.

**work process**—See *process*.

**yellow and red cards**—Signals for stopping people from talking too much, going beyond the scope of the discussion, or in other ways obstructing the group's work.

# Index

# Mapping Work Processes
## Second Edition
Bjørn Andersen, Tom Fagerhaug, Bjørnar Henriksen, and Lars E. Onsøyen

This peerless best seller is a hands-on, step-by-step workbook of instructions on how to create flowcharts and document work processes. No other book even comes close in teaching practitioners these crucial techniques.

The most noticeable change in this second edition is the inclusion of several new types of process maps. While the basic, straightforward flowchart is still extensively used, it has been supplemented by a number of other types, all of which serve different purposes. The authors have therefore expanded the variety of charts taught.

All the mapping techniques have also been updated, the mapping exercise is put into a larger context, and organizational examples from many different industries are used throughout to help readers understand real-life applications of the material presented. Also new is an example case study carried throughout the entire book to illustrate the construction and use of the different types of process maps.

**About the Authors:** **Bjørn Andersen** is a professor at the Norwegian University of Science and Technology, where he teaches business process and performance management. He also holds a position as research director at the research foundation SINTEF, where he heads a group of researchers within project and performance management. He is the author of several books published by ASQ Quality Press, in addition to papers in international journals and conferences. Andersen holds a master's degree in industrial engineering from the Norwegian Institute of Technology and a Ph.D. in benchmarking from the same university, including a year's stay at Rochester Institute of Technology. He is a Senior member of ASQ and associate member of the International Academy of Quality.

**Tom Fagerhaug** is an associate professor at the Norwegian University of Science and Technology, where he teaches quality, performance, and operations management. He also holds a position as research scientist at SINTEF, where he works on business process improvement and performance measurement projects. He is the author of several books published by ASQ Quality Press, in addition to numerous papers in international journals and at conferences. Fagerhaug holds a master's degree in industrial engineering from the Norwegian Institute of Technology and Ph.D. in quality management from the Norwegian University of Science and Technology. He is a Senior member of ASQ.

**Bjørnar Henriksen** is a senior research scientist at SINTEF, where he is involved in contract research within production strategy, process improvement, and performance management. Previously Henriksen worked as a project manager for Cap Gemini. He also has a broad consulting experience from other consulting and advisory firms, such as Ernst & Young. Henriksen holds an MSc in business from the University of Agder, Norway, and is currently pursuing a Ph.D. at the Norwegian University of Science and Technology in the field of production strategy.

**Lars E. Onsøyen** is a research scientist at SINTEF, where he is involved in contract research within business process improvement and performance management. Previously Onsøyen worked as a research officer at Stavanger University College. He has also held positions as a naval officer and as organizational development of the Directorate of the Norwegian Labor Inspection Authority. Onsøyen holds an MSc in org from the London School of Economics and Political Science (LSE).

Quality Press
600 N. Plankinton Avenue
Milwaukee, Wisconsin 53203
Call toll free 800-248-1946
Fax 414-272-1734
www.asq.org
http://www.asq.org/quality-press
http://standardsgroup.asq.org
E-mail: authors@asq.org

ASQ
AMERICAN SOCIETY
FOR QUALITY

H1170